FLASH FLOOD FORECASTING OVER COMPLEX TERRAIN

With an Assessment of the Sulphur Mountain NEXRAD in Southern California

Committee to Assess NEXRAD Flash Flood

Forecasting Capabilities at Sulphur Mountain, California

Board on Atmospheric Sciences and Climate

Division on Earth and Life Studies

NATIONAL RESEARCH COUNCIL
OF THE NATIONAL ACADEMIES

THE NATIONAL ACADEMIES PRESS
Washington, D.C.
www.nap.edu

THE NATIONAL ACADEMIES PRESS 500 Fifth Street, NW Washington, DC 20001

NOTICE: The project that is the subject of this report was approved by the Governing Board of the National Research Council, whose members are drawn from the councils of the National Academy of Sciences, the National Academy of Engineering, and the Institute of Medicine. The members of the committee responsible for the report were chosen for their special competences and with regard for appropriate balance.

Support for this project was provided by the National Oceanic and Atmospheric Administration under Contract No. 50-DGNA-1-90024. Any opinions, findings, and conclusions or recommendations expressed in this publication are those of the author(s) and do not necessarily reflect the views of the organizations or agencies that provided support for the project.

International Standard Book Number 0-309-09316-3 (Book)
International Standard Book Number 0-309-54580-3 (PDF)
Library of Congress Control Number 2004116335

Additional copies of this report are available from the National Academies Press, 500 Fifth Street, N.W., Lockbox 285, Washington, DC 20055; (800) 624-6242 or (202) 334-3313 (in the Washington metropolitan area); Internet, http://www.nap.edu.

Cover: Base (0.5°) reflectivity images as seen from the Sulphur Mountain NEXRAD (front) and the adjoining Santa Ana Mountains NEXRAD (back) on February 18, 2004 approximately two minutes apart. Images courtesy of WeatherTAP Inc.

Printed in the United States of America

THE NATIONAL ACADEMIES

Advisers to the Nation on Science, Engineering, and Medicine

The **National Academy of Sciences** is a private, nonprofit, self-perpetuating society of distinguished scholars engaged in scientific and engineering research, dedicated to the furtherance of science and technology and to their use for the general welfare. Upon the authority of the charter granted to it by the Congress in 1863, the Academy has a mandate that requires it to advise the federal government on scientific and technical matters. Dr. Bruce M. Alberts is president of the National Academy of Sciences.

The **National Academy of Engineering** was established in 1964, under the charter of the National Academy of Sciences, as a parallel organization of outstanding engineers. It is autonomous in its administration and in the selection of its members, sharing with the National Academy of Sciences the responsibility for advising the federal government. The National Academy of Engineering also sponsors engineering programs aimed at meeting national needs, encourages education and research, and recognizes the superior achievements of engineers. Dr. Wm. A. Wulf is president of the National Academy of Engineering.

The **Institute of Medicine** was established in 1970 by the National Academy of Sciences to secure the services of eminent members of appropriate professions in the examination of policy matters pertaining to the health of the public. The Institute acts under the responsibility given to the National Academy of Sciences by its congressional charter to be an adviser to the federal government and, upon its own initiative, to identify issues of medical care, research, and education. Dr. Harvey V. Fineberg is president of the Institute of Medicine.

The **National Research Council** was organized by the National Academy of Sciences in 1916 to associate the broad community of science and technology with the Academy's purposes of furthering knowledge and advising the federal government. Functioning in accordance with general policies determined by the Academy, the Council has become the principal operating agency of both the National Academy of Sciences and the National Academy of Engineering in providing services to the government, the public, and the scientific and engineering communities. The Council is administered jointly by both Academies and the Institute of Medicine. Dr. Bruce M. Alberts and Dr. Wm. A. Wulf are chair and vice chair, respectively, of the National Research Council.

www.national-academies.org

COMMITTEE TO ASSESS NEXRAD FLASH FLOOD FORECASTING CAPABILITIES AT SULPHUR MOUNTAIN, CALIFORNIA

PAUL L. SMITH, (*Chair*), South Dakota School of Mines and Technology, Rapid City
ANA P. BARROS, Duke University, Durham, North Carolina
V. CHANDRASEKAR, Colorado State University, Fort Collins
GREGORY S. FORBES, The Weather Channel, Inc., Atlanta, Georgia
EVE GRUNTFEST, University of Colorado, Colorado Springs
WITOLD F. KRAJEWSKI, University of Iowa, Iowa City
THOMAS D. POTTER, University of Utah, Salt Lake City
RITA ROBERTS, National Center for Atmospheric Research, Boulder, Colorado
MATTHIAS STEINER, Princeton University, New Jersey
ROGER M. WAKIMOTO, University of California, Los Angeles

NRC Staff

JULIE DEMUTH, Study Director
ELIZABETH A. GALINIS, Senior Program Assistant

BOARD ON ATMOSPHERIC SCIENCES AND CLIMATE

ROBERT J. SERAFIN *(Chair)*, National Center for Atmospheric Research, Boulder, Colorado
FREDERICK R. ANDERSON, McKenna Long and Aldridge LLP, Washington, D.C.
ROBERT C. BEARDSLEY, Woods Hole Oceanographic Institution, Massachusetts
ROSINA M. BIERBAUM, University of Michigan, Ann Arbor
RAFAEL L. BRAS, Massachusetts Institute of Technology, Cambridge
MARY ANNE CARROLL, University of Michigan, Ann Arbor
WALTER F. DABBERDT, Vaisala Inc., Boulder, Colorado
KERRY A. EMANUEL, Massachusetts Institute of Technology, Cambridge
CASSANDRA G. FESEN, Dartmouth College, Hanover, New Hampshire
JENNIFER A. LOGAN, Harvard University, Cambridge, Massachusetts
WILLIAM J. RANDEL, National Center for Atmospheric Research, Boulder, Colorado
ROGER M. WAKIMOTO, University of California, Los Angeles
JOHN C. WYNGAARD, Pennsylvania State University, University Park

Ex Officio Members

ANTONIO J. BUSALACCHI, JR., University of Maryland, College Park
ERIC F. WOOD, Princeton University, New Jersey

NRC Staff

CHRIS ELFRING, Director
AMANDA STAUDT, Senior Program Officer
JULIE DEMUTH, Program Officer
SHELDON DROBOT, Program Officer
ELIZABETH A. GALINIS, Senior Program Assistant
ROB GREENWAY, Senior Program Assistant
DIANE GUSTAFSON, Administrative Coordinator
PARIKHIT SINHA, Postdoctoral Fellow
ANDREAS SOHRE, Financial Associate

Preface

The request from the National Oceanic and Atmospheric Administration (NOAA) to the National Academy of Sciences (NAS) to conduct a study to examine the availability, performance, and capability of the Sulphur Mountain NEXRAD located in Ventura County, California, originated from U.S. Senator Barbara Boxer (California). In a letter to NAS President Bruce Alberts she explained that her motivations were based on the potential threat that flash floods pose to the residents of Ventura and Los Angeles Counties and concerns that the Sulphur Mountain NEXRAD is not effective in providing sufficient warnings (Appendix A). In addition, Senator Boxer specifically requested an examination of the warning failure rate and the radar data gap. Upon receipt of this request, the National Academies Board on Atmospheric Sciences and Climate (BASC) developed a charge that was agreed to by Senator Boxer, NOAA, and the National Academies Governing Board.

In January 2004, the Committee to Assess NEXRAD Flash Flood Forecasting Capabilities at Sulphur Mountain, California, was formed to undertake this study. Over the next six months, we proceeded to hold four meetings, three of which included information-gathering sessions. At the first meeting, we learned first-hand about the motivation for this study as well as the NOAA National Weather Service (NWS) modernization effort and its flash flood warning mission. The committee held its second meeting in Ventura, California, where it conducted two site visits—one to the Los Angeles-Oxnard NWS Weather Forecast Office (WFO) and one to the site of the Sulphur Mountain NEXRAD. In addition, a full day of that meeting was devoted to hearing nine invited presentations on topics ranging from flash flood hydrology to forecasting flash floods to technical issues associated with radars sited in complex terrain and near coastal regions. At the end of that same day, the committee opened the floor to the public, allowing anyone to make comments for consideration by the committee. During this open forum, some of the concerns expressed were with regard to potential health effects posed by the radar. However, the committee's interpretation is that the issue of health effects is not within the purview of the Statement of Task, which was agreed upon by both Senator Boxer and NOAA, and therefore it is not discussed in this report. Our public information-gathering efforts were concluded at our third meeting by two more experts who provided information about operational flash flood forecasting and the siting of the Sulphur Mountain radar.

The committee considered carefully all of the input and materials provided at these meetings, as well as statistics on nationwide flash flood warnings and events, the Sulphur Mountain radar availability, and atmospheric inversion data, which the committee specifically requested from NWS officials and others. After reviewing all of this information, the committee found little substantive basis for the concerns that had been expressed about the flash flood forecasting and warning capabilities of the Los Angeles-Oxnard WFO. Because there always is room for improvement, however, the committee developed its own analyses of the coverage provided by the Sulphur Mountain radar and determined ways in which that coverage could be improved. This report summarizes the information obtained throughout the study and presents the resulting findings and recommendations.

On behalf of the entire committee, I want to thank the many people who provided data, information, and opinions throughout the course of this study; the committee is solely responsible for its interpretations of all the input received. I especially want to thank Daniel Melendez at NWS Headquarters for fielding the numerous data requests of the committee; Dave Danielson and Dan Keeton at the Los Angeles-Oxnard WFO for their assistance with the Los Angeles-Oxnard data; and Michal Kraszewski, Radosław Goska, Anton Kruger, and Alexandros Ntelekos for providing us with detailed radar coverage and power loss calculations. I also want to express appreciation to Milton Kramer for his attention to this issue; not far into this study, the committee learned to appreciate the passion of the Ventura County people for this issue, and I think the committee would agree that this made for an interesting and informative study. Thank you also to Bob Serafin, our liaison to BASC, who attended a couple of our meetings and contributed many thoughtful ideas, and to Connie Crandall for her assistance in preparing some of our drafts of this report. And a special thank you to the BASC staff who assisted with this project, including our Study Director, Julie Demuth, and our Senior Program Assistant, Liz Galinis.

Finally, I want to commend the committee—a tremendous group of thoughtful, hard-working, talented people who volunteered generously their time in order to complete this study in a condensed time frame. Via our many contemplative and provocative conversations, we provide this assessment of using NEXRAD radars in complex terrain for flash flood forecasting with a special assessment of the Sulphur Mountain NEXRAD. We hope it proves useful to operational forecasters, researchers, policymakers, and stakeholders.

Paul L. Smith, *Chair*
Committee to Assess NEXRAD Flash Flood Forecasting Capabilities at Sulphur Mountain, California

Acknowledgments

This report has been reviewed in draft form by individuals chosen for their diverse perspectives and technical expertise, in accordance with procedures approved by the National Research Council's Report Review Committee. The purpose of this independent review is to provide candid and critical comments that will assist the institution in making its published report as sound as possible and to ensure that the report meets institutional standards for objectivity, evidence, and responsiveness to the study charge. The review comments and draft manuscript remain confidential to protect the integrity of the deliberative process. We wish to thank the following individuals for their review of this report:

Howard B. Bluestein, University of Oklahoma, Norman
Leon E. Borgman, University of Wyoming, Laramie
Elbert (Joe) W. Friday, Jr., University of Oklahoma, Norman
Stanley Q. Kidder, Colorado State University, Fort Collins
Clifford E. Mass, University of Washington, Seattle
Iraj Nasseri, Los Angeles County Department of Public Works, California
Dallas Raines, KABC-TV, Los Angeles, California
James A. Smith, Princeton University, New Jersey
Soroosh Sorooshian, University of California, Irvine
James Wilson, National Center for Atmospheric Research, Boulder, Colorado

Although the reviewers listed above have provided constructive comments and suggestions, they were not asked to endorse the report's conclusions or recommendations, nor did they see the final draft of the report before its release. The review of this report was overseen by Kuo-Nan Liou, University of California, Los Angeles, and George Frederick, Vaisala, Inc. Appointed by the National Research Council, they were responsible for making certain that an independent examination of this report was carried out in accordance with institutional procedures and that all review comments were carefully considered. Responsibility for the final content of this report rests entirely with the authoring committee and the institution.

Contents

EXECUTIVE SUMMARY 1

1 INTRODUCTION 9

2 FLASH FLOODS 12
The Physical Processes of Flash Floods, 13

3 THE NATIONAL WEATHER SERVICE AND FLASH FLOODS 26
The Modernization and Associated Restructuring of the National Weather Service, 27
The Process of Issuing Flash Flood Forecasts, Watches, and Warnings, 29

4 NEXRAD 35
The NEXRAD Network, 35
NEXRAD Scan Strategies, 40
Precipitation Estimation, 41
The Evolving NEXRAD System, 49
Siting of the NEXRADs, 51

5 OBSERVATIONAL CHALLENGES OF LOW-LEVEL RADAR COVERAGE IN COMPLEX TERRAIN AND COASTAL AREAS 59
Beam Visibility, 60
Propagation of Electromagnetic Radar Signals, 61

6 FLASH FLOOD WARNING PROCESS IN LOS ANGELES AND VENTURA COUNTIES 74
Rainfall Records for Los Angeles and Ventura Counties, 77
The NWS Los Angeles-Oxnard Flash Flood Warning Process, 82

Private Sector Role in Forecasts and Dissemination, 85
Stakeholders, 86
Need for Multiagency and Public Collaboration, 88

7 EVALUATION OF THE SULPHUR MOUNTAIN RADAR AND FLASH FLOOD WARNINGS IN LOS ANGELES AND VENTURA COUNTIES 91
Radar Coverage, 91
Availability of the Sulphur Mountain Radar, 100
Flash Flood Warning Statistics, 101

8 POTENTIAL IMPROVEMENTS IN FLASH FLOOD WARNINGS 117
NEXRAD and Other Radar Sources, 118
Modeling, Data Assimilation, and Decision Support Systems, 135

9 CONCLUDING THOUGHTS 146

REFERENCES 149

APPENDIXES

A Origin of the Study 169

B Characteristics of the NEXRAD Radar 172

C Chronology of the Siting of the Sulphur Mountain NEXRAD 175

D National Weather Service Flash Flood Verification Procedures 176

E Acronyms 180

F Community Participation 184

G Committee and Staff Biographies 186

Executive Summary

In early 2004, the Committee to Assess NEXRAD Flash Flood Forecasting Capabilities at Sulphur Mountain, California, was formed in response to a 2003 congressional request from U.S. Senator Barbara Boxer (California) and with support from the National Oceanic and Atmospheric Administration (NOAA). The committee's general task was to assess the effectiveness of operating Next Generation (weather) Radars (NEXRADs) in complex terrain to support the National Weather Service (NWS) in its task of forecasting heavy precipitation events and issuing flash flood forecasts, watches, and warnings. The committee conducted a specific analysis of the Sulphur Mountain NEXRAD located in Ventura County, California, and also considered how flash flood forecasting could be improved for other NEXRADs sited in complex terrain.

FINDINGS REGARDING THE SULPHUR MOUNTAIN RADAR AND THE LOS ANGELES-OXNARD WEATHER FORECAST OFFICE

After conducting a thorough analysis of the availability of the Sulphur Mountain NEXRAD, the coverage provided by the radar, and the performance of the Los Angeles-Oxnard (LOX) NWS Weather Forecast Office (WFO) as measured by the flash flood warning statistics, the committee found little basis for concerns regarding the operational effectiveness of the Sulphur Mountain radar. The committee's review of atmospheric inversion height statistics off the coast of Southern California and anomalous propagation effects on neighboring NEXRADs indicates that the current placement of the Sulphur Mountain radar above the mean inversion height was justified. In addition, the committee's examination of the radar availability shows that, from the time of the release in October 1998 of the General Accounting Office report about this radar through December 2003, the average availability of the Sulphur Mountain radar exceeded the availability requirements established by the NWS.

The committee performed its own calculations of low-level radar coverage using 30-m digital terrain elevation model data, considering the full extent of the radar beam (i.e., not just the axis, as in some prior studies) and assuming standard atmospheric propagation conditions. An integrated analysis of the coverage provided by the Sulphur Mountain NEXRAD and the adjoining Vandenberg Air Force Base, Santa Ana Mountain, Edwards Air Force Base, and San Diego NEXRADs reveals overlapping coverage for much of the area. However, there is an area southwest of the Sulphur Mountain radar over the Pacific Ocean that is covered exclusively by this radar. Low-level coverage of this area from which storms often approach is important for monitoring incoming storms and assessing their flash flood potential before they move onshore. Although the center of the Sulphur Mountain radar beam at the current minimum antenna elevation angle of 0.5° rises above 1.83-km (6000-ft) altitude beyond 75 km from the radar site, half of the radar beam is below that level at this range and has enabled detection of approaching storms. The committee did a hypothetical analysis with the radar antenna lowered to 0.0° and found that this would provide beam-axis coverage at or below 1.83 km (6000 ft) out to 125 km.

The committee examined the LOX WFO's flash flood warning statistics and found that it has an excellent record of issuing flash flood warnings. The proportion of flash flood events with advance warning and the average warning lead time more than doubled following the commissioning of NEXRADs nationwide, including the Sulphur Mountain radar. When the LOX statistics are compared with those of the other 115 WFOs throughout the continental United States, their percentage of flash flood events with advance warnings, at 79 percent, is better than the national average of 69 percent. LOX's ratio of flash flood events that were forecast but failed to materialize (called the false alarm ratio) is less than the national average.

The average lead time for flash flood warnings from LOX is less than the national average. The national statistics, however, are skewed by the large numbers of WFOs that forecast for regions with little to no relief in the surrounding topography. A more accurate assessment of the LOX WFO's performance is gained by comparing LOX with the 2004 goals of the NWS Western Region, which are established by considering the complex terrain in the western United States and thus the greater tendency for rapid onset of flash flooding. LOX's average lead time, as well as the percentage of forecast events and the false alarm ratio, is superior to the 2004 Western Region goals. Based on these findings, the committee concludes that the LOX office is performing its flash flood warning mission in an effective manner.

Overall, the committee finds that the Sulphur Mountain radar is appropriately sited to detect approaching storms while avoiding problems with anomalous propagation of the radar signals. The radar is amply functional and has provided crucial support to the Los Angeles-Oxnard forecasters in their mission to monitor, predict, and warn of precipitating events and flash floods.

RECOMMENDATIONS

Despite the committee's findings that the LOX office is accomplishing its mission and that the Sulphur Mountain radar is adequately sited and functional to fulfill its purpose, there are several ways in which flash flood forecasting and warning can be improved, not only in Southern California and in other regions where NEXRADs are sited in complex terrain, but throughout the country.

Enhancing Weather Radar Coverage

Providing broad low-level radar coverage, especially in complex terrain, presents a range of challenges. The coverage from radar sites at low levels is limited by Earth's curvature and often blocked in many directions by the surrounding terrain. Coverage from elevated sites, on the other hand, may not extend down into many valley areas. The latter problem is exacerbated by the current restriction of the NEXRAD system to minimum elevation (tilt) angles of 0.5°.

Recommendation: The NWS should improve nationwide NEXRAD coverage of low-level precipitation and wind, especially for elevated radar sites in complex terrain, through the adoption of a modified scan strategy that will allow scanning at lower elevation angles (Chapter 8[1]).

The use of lower, and perhaps even negative, elevation angles would allow monitoring of precipitation and wind at lower altitudes and, hence, would provide a more representative assessment of near-surface rainfall

[1] Chapter numbers at the end of each recommendation indicate where that recommendation and its supporting text can be found.

rates. Flexible selection of elevation angle steps would allow greater ability to avoid terrain blockage and to capture low-level meteorological phenomena. The NWS should make necessary hardware and software changes to the NEXRAD system to allow this type of modified scan strategy at the Sulphur Mountain site and other NEXRAD installations nationwide.

Recommendation: To extend radar coverage, all available regional real-time weather radar data should be made accessible to the NWS WFOs, including Federal Aviation Administration (FAA) and Department of Defense (DoD) NEXRAD radars; FAA Terminal Doppler Weather Radars (TDWRs) and other surveillance radars equipped to provide weather-echo data; local television station Doppler radars; and operational radars from other organizations (Chapter 8).

High-quality, real-time weather radar data are becoming more widespread in the United States. Although some of these data may not be of the same quality as NEXRAD data, they should be made accessible to NWS forecasters to increase the area and density of coverage of weather radar data—especially in regions of complex terrain—to improve forecasts, watches, and warnings. For radars located along the coast, this could provide a valuable extension of the offshore coverage. The high temporal and spatial resolution of the TDWR data should be of particular value in flash flood forecasting and warning for the major urban areas near which those radars are situated. Data formats and standards must be adhered to and efforts must be made to ensure data quality. This recommendation is consistent with previous National Research Council reports (NRC, 2002, 2003).

Recommendation: The NWS should consider augmenting the NEXRAD network with additional short-range radars to improve observation of low-level meteorological phenomena (Chapter 8).

The NEXRAD is designed for long-range coverage with a single-beam antenna that provides coverage of a large volume of the atmosphere. Although this is adequate for many situations, NEXRAD coverage at low altitudes far away from the radar can be insufficient due to Earth's curvature and terrain-induced blockage. Additional temporary local problems may arise, for example, as in areas affected by recent wildfires. Some of these problems can be resolved by deploying additional radars, which could be smaller, cheaper, and more easily (and even adaptively) deployable than NEXRAD. To maximize their use, these systems should be networked together, data formats standardized, and metadata established.

Improving the NWS Flash Flood Event Warning and Evaluation Processes

Flash floods are meteorological, hydrological, and hydraulic events that often have economic and societal implications. Therefore, an effective flash flood forecaster must adequately monitor and predict an event based on environmental factors and must communicate the warning in a manner that is useful to stakeholders. Evaluation procedures need to be refined to reflect better the skill and value of the warnings.

Recommendation: NWS Weather Forecast Offices nationwide, including the Los Angeles-Oxnard Weather Forecast Office, should continue to expand their collaborative efforts with key stakeholders (e.g., county, police, and emergency management officials) to enhance the effectiveness of flash flood forecasts, watches, and warnings (Chapter 6).

The committee commends the Los Angeles-Oxnard WFO on its efforts to communicate with local emergency management officials and the media about potential flash flood situations. However, the committee was informed that the criteria for issuing flash flood warnings are not always sufficient to address all the specific needs of a WFO's end-user community. Existing and new tools—such as geographic information systems (GIS) and the Flash Flood Monitoring and Prediction (FFMP) program, polarimetric radar, and ensemble probabilistic modeling—will make it possible to provide warnings with improved specificity. Moreover, forecasts formulated in a manner analogous to the NWS convective outlooks and with an event severity index analogous to the Fujita scale for tornadoes would provide useful indications of the magnitude of threats. Flash flood forecasts couched in probabilistic terms also would be a useful way to convey uncertainty information. Advanced technologies enable new ways of displaying and communicating individualized warnings to improve the efficiency of the warning process. In the spirit of the recent National Research Council report, *Fair Weather* (NRC, 2003), this should be done in cooperation with the private sector. In addition, in partnership with the media and emergency management officials, the NWS should continue to educate the public about the risks of flash flooding.

Recommendation: Evaluation of flash flood warnings should be based on their contributions to improved decision making and should employ metrics that take account of the magnitude and scale of the events and the

increasing specificity of the warnings. The NWS should improve the database of flash flood events and warnings to include more complete and accurate listings of both warnings and events (Chapter 7).

The traditional national measures for evaluating flash flood warnings should be augmented to better take into account the myriad needs of the local population at risk. For example, the value attributed to the lead time of a warning should take into account factors such as the response time of a basin. In addition, warnings indicating the magnitude of the threat and designating specific regions within county warning areas are more useful to both the intended responders and others within the same county who are not affected by the warning. Evaluation statistics such as the probability of detection (POD) and the false alarm ratio (FAR) could be stratified by the scale of the events. The warning and event databases should be adapted to capture the increased specificity of the warning capability, in contrast to the current practice of recording warnings and events at the county level, with little indication of magnitude and impact.

Establishing a reliable database of events for verification purposes involves well-recognized difficulties, but recording the warnings issued is (or should be) a straightforward process. Nevertheless, situations involving a series of overlapping warnings, or extensions of warnings, and possibly multiple flash flood events within a county appear not to be well represented in the national database. This results, at least partly, from the rules employed in establishing that database; such rules may be required to ensure uniformity that permits comparisons across the nation. However, a more comprehensive listing would be useful for detailed analyses.

Guiding Future Directions

A flash flood by definition is a rapidly evolving event. Radars can observe the evolving weather systems that cause flash floods as they move into areas of concern, and future NEXRAD enhancements will improve the quality of those observations. The measurements of accumulating precipitation alone may not provide the most effective warning capability, but techniques such as the Flash Flood Monitoring and Prediction program will provide improvements. There also are gains to be made from rapidly evolving capabilities for better short-term and fine-scale forecasting, using regional and local numerical models that ingest real-time observations. In addition to these near-term improvements, consideration of hydrologic factors should be an integral part of future radar siting.

Recommendation: To increase the accuracy and lead time of flash flood forecasts and warnings, the NWS should continue to implement new technologies and techniques including (a) the Flash Flood Monitoring and Prediction program at all Weather Forecast Offices, (b) polarimetric modifications to NEXRAD, (c) data assimilation systems that integrate radar and other operational datasets into coupled hydrometeorological and hydrological models, and (d) data fusion systems (Chapter 8).

Extensive opportunities exist for forecasters to take advantage of the rapid advancement of technology to improve forecasts, watches, and warnings. The FFMP system, which requires adaptation to the specific watersheds served by each WFO, would facilitate more specific flash flood warnings. In addition, as part of its new Advanced Hydrologic Prediction Services, the NWS is encouraged to continue its effort to develop and evaluate hydrologic and coupled meteorological-hydrologic models to advance technologies useful for improved flash flood guidance and warnings. The polarimetric modification would improve the data quality and quantitative precipitation measurement capabilities of NEXRAD. Real-time data assimilation systems that incorporate observations into high-resolution mesoscale numerical models provide rapidly updated wind and precipitation forecasts. Data fusion systems, such as the Auto-Nowcast system, incorporate all available observation datasets together with numerical model output to produce very short range (0- to 2-hour), site-specific forecasts. These advances can produce improved forecasts, including ensemble and probabilistic forecasts, of precipitation rate and accumulation that are essential for flash flood forecasting. To enhance the usefulness of the forecast, quantification of uncertainty (e.g., probability forecasts) should be an integral component of these products (NRC, 2003).

Recommendation: In addition to the original NEXRAD siting considerations, future siting of radars in complex terrain should include detailed assessments of coverage in areas at risk for flash flooding (Chapter 8).

The original NEXRAD siting procedures considered primarily meteorological processes, radar coverage, and ground clutter. Increased understanding of hydrologic processes of runoff production and streamflow response, combined with the application of radar to real-time hydrologic prediction for individual catchments (e.g., FFMP), enables incorporation of hydrologic aspects of the land surface into future siting processes. Readily available detailed digital information on topography and other relevant spatial information now make it possible to analyze a radar's coverage of the near-ground portion of the atmosphere using GIS technology. The potential for complete and partial radar beam blockage can be evaluated in the context of hydrologic basins for which coverage is sought. Basin size, average slope, orientation with respect to the movement of dominant weather patterns, characteristics of soil, land cover and land use, and channel hydraulic aspects determine the amount of rain that is likely to cause flash flooding and where it may occur. These characteristics, together with local hazard vulnerabilities, can help determine priorities of site selection.

1

Introduction

Every year, floods are responsible for more deaths nationwide than any other weather phenomenon, with a 30-year average of more than 120 fatalities per year; of these, most are due to flash floods (NWS, 2004a). From 1996 through 2003, the average number of flash flood events recorded nationwide was nearly 3000 per year (NWS, 2004b). In comparison, only about 1000 tornadoes and 60 related deaths occur in the United States annually (Storm Prediction Center, 2004). Although Davis (2001) and the American Meteorological Society (BAMS, 2000) note that there have been recent improvements in flash flood forecasting due to better radar and satellite observations, flash floods continue to be among nature's worst killers. Forecasting flash floods is especially challenging because a forecaster must predict not only the occurrence of the event but also its magnitude, for it is the amount of precipitation and the time frame in which it occurs that can transform an ordinary rainfall event into a deadly one. The challenge is particularly great in regions with complex terrain and in which storms approach over water where there are sparse, intermittent surface-level observations.

The National Oceanic and Atmospheric Administration's (NOAA) National Weather Service (NWS) is the government organization charged with protecting the nation from extreme weather events, including flash floods. The NWS has more than 120 Weather Forecast Offices (WFOs), each of which is responsible for a designated group of county warning areas. Combined, the WFOs are responsible for monitoring the entire United States and its territories. The WFOs have sole official authority to issue flash flood watches and warnings. Their forecast process is extensive, but the crux of flash flood forecasting lies in monitoring storm formation, movement, and precipitation patterns.

In the mid-1990s, a nationwide network of Weather Surveillance Radar-1988 Doppler (WSR-88D) radars was established. Termed the Next

Generation Radar, or NEXRAD, this radar was a significant improvement over the previous NWS weather radars introduced in 1957 and 1974. There are more than 130 NEXRADs in the United States; with its increased coverage, denser network, greater spatial and temporal resolution, and improved precipitation products, NEXRAD became the major foundation for modern short-term precipitation forecasting. Nonetheless, questions exist about the efficacy of NEXRADs sited in complex terrain because of the potential for (1) gaps in radar coverage due to beam blockage by higher surrounding terrain and (2) lack of radar coverage at low levels of the atmosphere if the radar is sited at a higher altitude relative to surrounding terrain.

In particular, the Sulphur Mountain NEXRAD located north of Los Angeles in Ventura County, California, has been a subject of controversy in terms of its ability to detect precipitation events below 1.83-km (6000-ft) altitude and its use to assist local NWS meteorologists in forecasting and warning of flash flood events. Many analyses of the Sulphur Mountain radar have been conducted (e.g., Paris 1997a, 1997b, 1998, 2001; Rose Institute, 1997, 1998; GAO, 1998; NWS, 2001a; Thompson, 2001; Fox, 2003), but questions persisted about its usefulness. Consequently, in February 2003, an appropriations bill was passed containing a clause—based on a request from U.S. Senator Barbara Boxer—directing that NOAA commission the National Academy of Sciences to conduct an objective study to assess the availability, performance, and capability of the Sulphur Mountain radar to detect heavy precipitation and aid forecasters at the NWS's Los Angeles-Oxnard WFO in providing flash flood forecasts and warnings (Appendix A). The specific charge to the study committee was to

- describe the overall strategy of the NEXRAD radars in support of the NWS flash flood warning and forecast mission and discuss strengths and weaknesses of the system for operations in complex terrain;
- assess the availability, performance, and capability of the NWS NEXRAD located on Sulphur Mountain in Ventura County, California, to detect heavy precipitation and aid forecasters at the Los Angeles WFO in providing flash flood warnings and forecasts and fulfilling its other intended purposes;
- assess how the Sulphur Mountain NEXRAD radar's location affects its capability to detect low-level storm events (i.e., below 6000 ft);
- provide conclusions about strengths and weaknesses and make recommendations to improve the accuracy and timeliness of flash flood warnings in and around western Los Angeles and Ventura Counties, California,

including any plausible alternative approaches for flash flood forecasting; and

- identify lessons that could benefit forecasters, facility planners, and decision makers as they deal with other NEXRAD installations and similar instruments deployed in the future.

The Committee to Assess NEXRAD Flash Flood Forecasting Capabilities at Sulphur Mountain, California, carried out this requested study during the first half of 2004. In doing so, the committee assessed the Sulphur Mountain radar and its availability, coverage, and use in flash flood forecasting; it also considered how this example could be applied to other NEXRADs located in regions of complex terrain for improved flash flood forecasting and warning. This report is the result of the considerations and deliberations undertaken by the committee.

For the interested reader, Chapters 2 through 4 provide technical background information relevant to this study. Chapter 2 provides basic information about the physical processes of flash floods. Chapter 3 provides an overview of the NWS, its modernization effort, and the general process undertaken by NWS forecasters to issue flash flood forecasts, watches, and warnings. An extensive description of NEXRAD is presented in Chapter 4, including discussions of the radar characteristics and the network and its use for precipitation estimation.

The reader more interested in the committee's analysis, findings, and recommendations should see Chapters 5 through 9. Chapter 5 presents special challenges of low-level radar coverage in regions of complex terrain. Chapters 6 and 7 look specifically at Los Angeles and Ventura Counties, with the warning process presented in Chapter 6 and the analysis of the Sulphur Mountain radar and the flash flood warning performance of the Los Angeles-Oxnard NWS office in Chapter 7. Chapter 8 outlines potential improvements in flash flood warning capabilities, and finally, Chapter 9 provides the concluding thoughts of the committee.

2

Flash Floods

> A flash flood is a flood that rises and falls quite rapidly with little or no advance warning, usually the result of intense rainfall over a relatively small area.[1]

Most flash floods in the United States are caused by intense rainfall from slow-moving thunderstorms, thunderstorms that repeatedly move over the same location, or excessive rainfall from hurricanes or other tropical systems (Junker, 1992; Davis, 2001; Kelsch, 2001). Along the West Coast of the United States, in contrast, flash floods frequently are caused by orographic precipitation—that is, precipitation influenced by mountainous terrain (Maddox et al., 1980; Cotton and Anthes, 1989; R. Smith et al., 1997). There, flash floods are typically associated with land-falling extratropical cyclones and fronts during the winter months, rather than summer thunderstorms. This leads to somewhat different forecasting considerations and poses challenges in measuring storm rainfall along the West Coast, as discussed throughout this report. Moreover, flash flooding is determined not only by meteorological factors but also by hydrological factors—such as terrain slope, land use, vegetation and soil types, and soil moisture—and by hydraulic processes related to the character of stream or river channels subject to flooding. For example, various combinations of rainfall intensity and duration may lead to flash flooding, depending on the hydrologic and hydraulic factors of a watershed. Flash floods result from a complex interaction among hydrometeorological, hydrological, and hydraulic processes across various spatial and temporal scales.

[1]Definition of a flash flood from the American Meteorological Society *Glossary of Meteorology* (AMS, 2000).

THE PHYSICAL PROCESSES OF FLASH FLOODS

The National Weather Service (NWS) Hydrologic Glossary (NWS, 2004c) defines flooding as the inundation of a normally dry area caused by high flow or overflow of water in an established water channel (e.g., river, stream, drainage ditch) or the ponding of water at or near the location where substantial rain fell. Flooding occurs whenever a drainage system receives more water than it can handle. As noted at the beginning of this chapter, a flash flood is a flood that follows the causative event in a short period of time and often is characterized by a sudden increase in level and velocity of a flowing water body. The term "flash" reflects a rapid response to the causative event, with rising water levels in the drainage network reaching a crest within minutes to a few hours of the onset of the event, leaving extremely short time for warning. A threshold of approximately 6 hours often is employed to distinguish a flash flood from a slow-rising flood (Mogil et al., 1978; Georgakakos, 1986a; Gruntfest and Huber, 1991; Polger et al., 1994; NWS, 2004c). Thus, flash floods are localized phenomena that occur in watersheds with maximum response times of a few hours—that is, at spatial scales of approximately 10,000 km^2 or less, depending on the catchment characteristics (Hirschboeck, 1987; Gruntfest and Huber, 1991; O'Conner and Costa, 2004). Most flash floods occur in streams and small river basins with a drainage area of a few hundred square kilometers or less (Kelsch, 2001). Such basins respond rapidly to intense rainfall rates because of steep slopes and impermeable surfaces, saturated soils, or because of human- (i.e., urbanization) or fire-induced alterations to the natural drainage.

Causative events may be either excessive rainfall in a natural drainage basin or human-altered catchment or the sudden release of water impounded by a natural jam (i.e., formed by ice or rock, mud, and wood debris) or human-made dam or levee. This report focuses on flash flood events associated with heavy rainfall. Extraordinary unit discharges (i.e., the rate of water flowing past a stream gauging station divided by the drainage area) from a watershed are related to specific topographic and climatologic conditions (Kelsch, 2001; O'Conner and Costa, 2004). In general, basins producing high unit discharges correspond to areas in which regional climatic patterns can produce extraordinary precipitation, such as within and flanking the Appalachian Mountains along the Atlantic seaboard, the western or southwestern flanks of mountain ranges near the Pacific Coast, and a broad northeast-trending zone in the southern Midwest extending from southwest Texas to southeast Kansas and southern Missouri. Such extraordinary rainfall situations feed off moisture sources originating from the Gulf of Mexico and the Pacific or Atlantic Oceans. In addition to the natural causes of flash floods, however, the increasing influence of urbanization cannot be ignored. The following sections cover aspects of the hydrometeorology, hydrology, and hydraulics of flash floods, complemented by a discussion of the human impacts of urbanization.

Hydrometeorological Processes

Flash floods typically result from intense rainfall rates that occur in individual thunderstorms, or lines or clusters of thunderstorms, or in conjunction with bands of heavy rain and showers in tropical or extratropical cyclones (Schwarz, 1970; Maddox et al., 1978, 1979; Caracena et al., 1979; Chappell, 1986; J. A. Smith et al., 1996; Landel et al., 1999; Hjelmfelt, 1999; J. A. Smith et al., 2000; Sturdevant-Rees et al., 2001). Most extreme precipitation events are associated with mesoscale weather systems—that is, systems with horizontal scales of 10 to 1000 km. These include individual thunderstorms on scales of a few kilometers to tens of kilometers and clusters of thunderstorms (sometimes called mesoscale convective systems, or MCSs) on scales of a few hundred kilometers. Precipitation often varies significantly on spatial scales of a few kilometers, especially during extreme events. Moreover, complex topography can exaggerate the spatial variability of rainfall.

There are several common meteorological scenarios conducive to a prolonged duration of localized heavy rainfall (Maddox et al., 1979; Doswell et al., 1996; Kelsch, 2001; Lin et al., 2001). The key ingredients of flash flood-producing storms are (1) ample and persistent supply of water vapor, (2) a mechanism to facilitate uplift of air in which the moisture condenses and precipitation forms, and (3) a focusing mechanism (or combination of focusing mechanisms) that causes precipitation to occur continuously or repeatedly over the same area. The physical processes at work are of dynamical, thermodynamical, and microphysical nature. The dynamic and thermodynamic conditions of the atmosphere determine the microphysical processes and the associated efficiency of precipitation formation and growth. Precipitation processes, however, have a significant effect on the dynamic and thermodynamic structure also. These feedback mechanisms of moist processes are only starting to be understood (Barros and Kuligowski, 1998; Rotunno and Ferretti, 2001; Colle, 2004).

The uplift of air may result from convective instability (i.e., the presence of warm, moist air at low levels and cooler air aloft), but often it is augmented or triggered by a focusing mechanism such as a front or mesoscale boundary (i.e., a smaller-scale front associated with, for example, the cool outflow from a previous or nearby thunderstorm or along a shoreline). The uplift may also be due to orography (e.g., moist air ascending a hillside). Large-scale ascent and bands of heavy showers and thunderstorms can also occur in association with extratropical or tropical storms. The greatest threat of excessive rainfall and significant flash flooding tends to occur when and where thunderstorms are slow-moving or stationary, continually reform over the same area, or repeatedly move over the same location. For example, if new convection develops on the storm flank opposite the direction of thunderstorm motion, then a quasi-stationary storm complex may evolve (J. A. Smith et al., 1996, 2000). On the other hand, storm movement may limit the duration of extreme rainfall rates at any given point and yield only moderate local rainfall accumulations.

Regional- and drainage basin-scale differences in flash flood-producing events across the United States are summarized by O'Conner and Costa (2004). For basins with areas greater than about 500 km^2, the highest unit discharges occur where large storms produce sustained rainfall for multiple days over broad areas, such as tropical storms in the eastern United States (Sturdevant-Rees et al., 2001) and Pacific cyclonic systems on the West Coast (Colle and Mass, 2000; Neiman et al., 2004). The highest unit discharges for smaller basins occur when convective systems deliver intense rainfall for periods of minutes to several hours. Such events are most common in Texas, the southeastern United States, and along the flanks of the Appalachians, where thunderstorms and mesoscale convective systems have produced near-world-record precipitation rates for durations up to 4 hours (J. A. Smith et al., 1996, 2000, 2001).

Complex Terrain

Special circumstances exist in complex terrain where a synoptically forced flow toward and over a topographic barrier may interact with the storm dynamics. This may lead to persistent (slow-moving or quasi-stationary) and orographically enhanced storm systems that produce heavy rainfall through both increased rain rates and increased time raining over a given area (R. B. Smith, 1979; Bruintjes et al., 1994; Buzzi et al., 1998; Rotunno and Ferretti, 2001; Medina and Houze, 2003). Often such precipitation systems are fed by a boundary layer jet (influenced by a stalled frontal boundary) that pumps near-saturated air into the storm, thereby facilitating an efficient formation of precipitation through predominantly warm rain processes (Box 2.1) at relatively low levels in the storm (Maddox et al., 1978; Caracena et al., 1979; Reinking and Boatman, 1986; J. A. Smith et al., 1996; Petersen et al., 1999; J. A. Smith et al., 2000; Kelsch, 2001). This low-level maximum in storm intensity poses a real challenge for radar detection in complex terrain or at farther ranges, where the lowest radar beam may be blocked or overshooting the low-echo-centroid signature (Wilson and Pollock, 1974; Westrick et al., 1999; White et al., 2003). In addition, melting snow layers concurrent with intense or persistent rainfall in complex terrain may substantially increase the threat of flooding (Barros and Kuligowski, 1998).

BOX 2.1

Warm Rain Processes

According to Rogers (1979), most of the world's precipitation falls to the ground as rain, much of which is produced by clouds whose tops do not extend to temperatures less than 0°C. The mechanism responsible for precipitation in these "warm" clouds is coalescence among cloud droplets. By far the dominant precipitation-forming process in the tropics, coalescence also plays a role in midlatitude clouds whose tops may extend to subfreezing temperatures.

Cloud droplets are formed through a process called nucleation, when water vapor condenses onto small hydrophilic aerosol particles. These tiny cloud droplets grow in size as additional water vapor is deposited onto them, a process called diffusional growth. Very small cloud particles have negligible fall velocities; however, as they grow in size their fall speeds increase as well. Differences in fall speed lead to collisions among the cloud particles, and some of these collisions result in combined, larger particles. This process of collisions among and coalescence of cloud droplets is a much more effective growth process than growth through deposition of water vapor; thus, raindrop-sized precipitation particles may rapidly develop. Moreover, the larger raindrops interact with cloud droplets as well as other raindrops, and this "warm rain process" of collision and coalescence continues. Growing particles may ultimately reach a size at which the surface tension is no longer able to keep the body of water together, and a large raindrop may break up into a number of smaller ones. Breakup of large raindrops may also be caused by collisions with other particles. Other factors limiting the development of rain are the supply of water vapor and the number of available hydrophilic aerosol particles for cloud droplet nucleation.

Regions with an ample supply of moisture from large water bodies, whether seasonally or throughout the year, may exhibit a higher fraction of surface rainfall produced by the warm rain process (as contrasted with the "cold rain" process, which involves evolution from ice particles in the clouds). If this supply of water vapor is combined with orographic uplift (e.g., along the California coastal mountains), this may lead to very effective precipitation production and flash flooding situations.

The coastal mountains, extending along virtually the entire length of western North America, represent a significant barrier to the lower tropospheric flow (Bond et al., 1997). These mountains can provide lift of moisture-laden air even under stably stratified conditions (Maddox et al., 1980). Indeed, substantial rainfall often follows when sustained strong winds associated with extratropical cyclones force high-humidity air up the mountains of California and the Pacific Northwest. Sometimes this deep plume of moisture originates from the subtropics near Hawaii; thus, the term "pineapple express" was coined.

Heavy rainfall in California and the Pacific Northwest is not always confined to the immediate west slopes of the coastal mountains. Often, a surface layer of cooler, stable air may exist upstream of the mountains prior to the onset of a precipitation event. In these situations, a strong southeasterly flow of air, called a "barrier jet," may form parallel to and west of the

---

Experimental Studies of Orographic Precipitation

The past few decades have seen a number of concerted, experimental efforts to advance a conceptual and quantitative understanding of the effects of terrain on the flow of moist air past it and the generation of precipitation on the upstream side. Examples along the mountain ranges of the U.S. Pacific northwest coast are the Cyclonic Extratropical Storms Project (CYCLES; Hobbs et al., 1975; Houze et al., 1976; Matejka et al., 1980; Hobbs and Persson, 1982); the Coastal Observation and Simulation with Topography Experiment (COAST; Colle and Mass, 1996; Bond et al., 1997; Braun et al., 1997); and most recently the Improvement of Micro physical Parameterization Through Observational Verification Experiment (IMPROVE; Stoelinga et al., 2003). Details about IMPROVE may be obtained from *http://improve.atmos.washington.edu*; moreover, a special issue of the *Journal of the Atmospheric Sciences* is in preparation. Further to the south, the Sierra Cooperative Pilot Project (SCPP; Marwitz, 1983; Reynolds and Dennis, 1986; Reynolds and Kuciauskas, 1988; Pandey et al., 1999) was conducted over the Sierra Nevada Mountains. In addition, California has seen numerous recent efforts to study orographic precipitation by means of the California Landfalling Jets Experiment (CALJET; Neiman et al., 2002, 2004; Ralph et al., 2003, 2004; White et al., 2003), the Pacific Jets Experiment (PACJET), and the National Oceanic and Atmospheric Administration Hydrometeorological Testbed (HMT). Particulars about these recent and ongoing projects may be found at *http://www.etl.noaa.gov/programs*. Another notable, internationally coordinated experimental effort, the Mesoscale Alpine Program (MAP; Bougeault et al., 2001), was focused on the European Alps. Information about this project may be retrieved from *http://www.map.ethz*.ch and from the many articles compiled in a special issue of the *Quarterly Journal of the Royal Meteorological Society* (QJRMS, 2003).

---

mountains. Thus, the approaching weather system with its strong west or southwest winds encounters this trapped layer near the coastline, forcing the moist air to rise offshore. Consequently, the heaviest precipitation may fall west of the mountain slopes, above the cold layer of blocked or trapped flow (Neiman et al., 2002). The presence of the barrier jet complicates the forecast and warning process, because it can erode (i.e., evaporate) rainfall otherwise headed toward the mountain slopes.

In some parts of the country, including the Rocky and the Appalachian Mountains, flooding due to orographic precipitation is from deep convection, as thunderstorms remain stationary or continually form in air lifted over the sloping terrain. In many of these areas, moist winds blow in at low levels from Gulf of Mexico and Atlantic Ocean moisture sources to the east or southeast, and the dangers are greatest in the warm season. Sometimes the moist air in these instances has its origin behind, or north of, a weak cold front. Although thunderstorms in which ice processes tend to dominate the development of precipitation have been involved in the worst storms in these areas, warm rain processes have been found to make important contributions to rainfall generation (Maddox et al., 1978; Reinking and Boatman, 1986).

Southern California

California's coastal mountain ranges are typically less than 100 km long and extend vertically only about 500–1500 m above mean sea level, but they can generate significant, orographically induced floods (Neiman et al., 2002). These events usually involve relatively shallow bands of rain and showers associated with land-falling extratropical cyclones and fronts in the winter months. White et al. (2003) show that the radar echoes often top around 3 km or less above mean sea level. In these cases, a strong south or southwesterly moist low-level flow from the Pacific Ocean impinges on the upward-sloping terrain. Despite a low-lying freezing level (around 2 km) during the winter months, the coastal convergence and orographic uplift may still result in significant concentrations of liquid water within the warm cloud layer. Thus, these storms can be associated with greater than 200 mm (8 in.) of rain in 24 hours and strong surface winds in excess of 50 m s^{-1}. For example, Davis (2001) noted that on January 3–5, 1982, upslope winds produced widespread rainfall near Santa Cruz, California, with rainfall amounts of 610 mm (24 in.) accumulated over a period of 28 to 30 hours in the higher elevations (National Climatic Data Center [NCDC], 1982). Although much of the precipitation can be attributed to orographically enhanced widespread (stratiform) rainfall, potential instability may be present and realized when air parcels are pushed up the terrain (Lowndes, 1968; Browning et al., 1974; Neiman et al., 2002). This allows embedded showers, and sometimes thunderstorms, to develop and locally increase rainfall rates. Because of the high population density, flooding along California's coastal mountains may produce millions of dollars in property damage and fatalities (NCDC, 1995, 1998).

White et al. (2003) suggest that nearly one-quarter of the total rainfall at a California coastal site may be attributed to warm rain processes. Although these events tend to be associated with light rain (less than approximately 5 mm h^{-1} [0.2 in. h^{-1}]), they can reach rain rates of 12–18 mm h^{-1} (0.5–0.7 in. h^{-1}). Such values would not normally cause flooding in most other locations across the country; for the Los Angeles area, however, rain rates of 12 mm h^{-1} (0.5 in. h^{-1}) or more surpass the rule of thumb used by local forecasters for guidance in issuing flood statements.

A number of modeling studies have shown the dependence of orographic precipitation on the upslope flow (Collier, 1975; Bell, 1978; Sinclair, 1994). Browning et al. (1975) and Neiman et al. (2004) show that the low-level jet (LLJ) region affecting the windward slopes of coastal mountains ahead of land-falling cold fronts within extratropical cyclones can be a critical contribution to orographic precipitation. In a seminal paper on flash flooding along the California coastal range, Neiman et al. (2002) quantitatively show that the amount of precipitation that falls on the coastal ranges of California is related to the magnitude of the upslope flow. They calculated vertical profiles of the linear correlation coefficient of upslope flow versus rain rate for all cases containing an LLJ in the winter season 1997–1998. These correlation coefficient profiles show a direct relationship between the magnitude of the upslope component of the LLJ flow affecting the coast and the rain rate in the downstream coastal mountains. Maximum correlation coefficients were as large as 0.94 in some individual cases and 0.70 for the entire winter season. Neiman et al. (2002) also found that the layer of upslope flow that optimally modulates orographic rainfall is near the mountaintop, approximately 1 km above mean sea level for California's coastal range. This height also corresponds to the mean altitude of the landfalling LLJs.

Although the most common flash flood scenario for the Los Angeles area is based on winter orographic precipitation involving southerly or southwesterly surface winds, as described above, flooding problems occasionally arise in other contexts as well. For example, after the passage of winter storm cold fronts, elevated cold air moving in from the northwest may result in sufficient instability to trigger severe thunderstorms with hail, damaging winds, and occasionally brief torrential rains sufficient to result in at least urban flooding. Alternatively, moisture flowing northward from tropical weather systems west of Mexico may also lead to flooding. Moisture flowing northwestward from the Gulf of California or during the Southwest monsoon in late summer and fall can lead to flooding from slow-moving thunderstorms in the eastern parts of the Los Angeles-Oxnard forecast area of responsibility (i.e., over east slopes and to the east of the coastal mountains).

Global-scale climatic anomalies may exert a strong influence on the occurrence of severe regional winter floods in the southwestern United States. For example, Andrews et al. (2004) investigated the influence of the El Niño-Southern Oscillation (ENSO) on flood frequency and magnitude along the California coast. They showed that floods in coastal basins of Southern California (south of 35° N) are significantly larger during El Niño than La Niña periods (Figure 2.1). This is a consequence of increased precipitation during El Niño-like conditions, which is tied to a frequent southward displacement and slight rotation of the wintertime polar jet to result in more southerly flow along California (see also Mitchell and Blier, 1997; Cayan et al., 1999; Masutani and Leetmaa, 1999). Such systematic shifts in wind direction relative to the coastal mountain orientation have an impact on the upslope flow and the corresponding "rain shadows" (Ralph et al., 2003).

Analyses of California rainfall (Monteverdi and Null, 1997) reveal that El Niño and La Niña are not the only factors that influence annual variations in rainfall and flooding in California, however. Although every Type I El Niño since 1949 has resulted in above-average winter precipitation at the Los Angeles Civic Center, not every El Niño episode results in unusual flooding in Southern California (Monteverdi and Null, 1997). Moreover, many flood events occur without the presence of El Niño. Of California's ten costliest floods since 1949, only three occurred during El Niño episodes (Null, 2001).

Hydrological Processes

Steep orography can funnel the excessive runoff from heavy rains into river basins, increasing the chance for flooding conditions and underscoring the fact that flash floods are not caused solely by meteorological phenomena. Heavy rainfall is required, but flooding also depends on the hydrologic characteristics of the watershed in which rainfall accumulates. The amount of runoff and the magnitude of streamflow depend on the rainfall distribu-

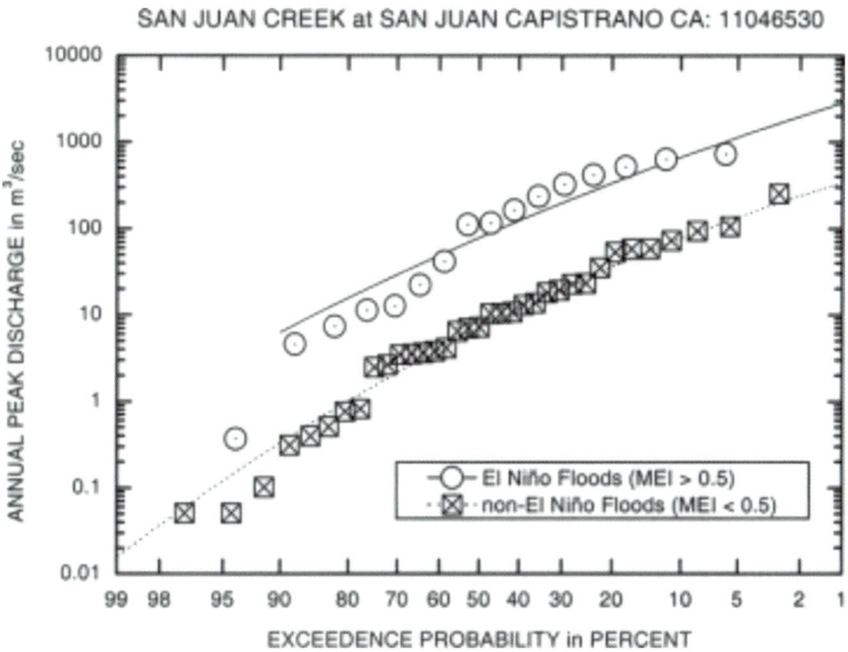

FIGURE 2.1 Comparison of observed El Niño and non-El Niño (i.e., neutral and La Niña) annual peak floods with fitted three-parameter log gamma distribution computed for San Juan Creek (U.S. Geological Survey station number 11046530) at San Juan Capistrano draining an area of 282 km^2. SOURCE: Figure 5 of Andrews et al. (2004).

tion, rainfall rate, and duration of high rainfall rates in the drainage basin. Physical characteristics of the basin (e.g., soil saturation and permeability, geography or slope, urbanization, vegetation) further modulate the runoff potential (Davis, 2001).

Hydrologic processes determine what happens to water reaching the land surface. Raindrops falling toward the ground may be intercepted by vegetation such as trees, bushes, or other plants. Alternatively, raindrops may fall on open water bodies (e.g., lakes, rivers), impermeable surfaces (e.g., rooftops of buildings, roads, parking lots), or reach a land surface that absorbs them. The land surface may be regarded as a "hydrologic switch" that either directs rainwater into the soil layers (i.e., infiltration) or lets it run off the surface (i.e., overland flow). Ponding and surface runoff will occur whenever the land surface receives more rainfall than it can absorb, either naturally or due to man-made interference such as sealed surfaces. There are two types of natural processes that may generate excess rainfall runoff: infiltration excess and saturation excess (Beven, 2001). Infiltration excess is driven by the rainfall rate, while saturation excess tends to result from high rain volume. These processes are highly nonlinear and depend heavily on soil properties (e.g., saturated hydraulic conductivity, moisture deficit), which are quite variable in space and time, depending on land use and land cover conditions.

Across a catchment the saturated hydraulic conductivity may vary from essentially zero for impervious surfaces to approximately 600 mm h^{-1} (24 in. h^{-1}) for sandy soils (Clapp and Hornberger, 1978). Common soil types, such as silty clay loam soils, may absorb rainfall rates on the order of 5–10 mm h^{-1} (0.2–0.4 in h^{-1}) or less. The actual infiltration capacity of a soil, however, depends also on the degree of saturation of the soil layers with depth and the underlying water table (Sturdevant-Rees et al., 2001). Extended periods of rainfall over a given area may decrease the soil's infiltration capacity to absorb water and thus cause saturation of the upper soil layers, which increases the effective impervious area of a drainage basin. Similarly, frozen soil conditions or layers of snow covering the land surface may also contribute to essentially impervious areas. In addition, wildfires may dramatically alter portions of a natural watershed's surface through the production of water-repellent soil conditions that result from the combustion of vegetation and other organic matter, thus increasing the effectively impervious area. For example, the flood of July 13, 1996, that occurred in Buffalo Creek, Colorado, was due partly to a wildfire that had burned a significant portion of the watershed only 2 months earlier (Warner et al., 2000; Chen et al., 2001). The resulting drastically increased ratio of precipitation runoff to infiltration exacerbated the high rainfall amounts, leading to a deadly and destructive flash flood. The burning of trees also affects the precipitation interception capacity. Both effects can significantly increase the runoff generation in the burned areas of a watershed (Johnson, 2000; Yates et al., 2001).

The temporal and spatial variability of rainfall appears to matter more for infiltration excess than for saturation excess runoff production (Winchel et al., 1998). For extreme events such as the Fort Collins, Colorado, flash flood on July 28, 1997, Ogden et al. (2000) found that uncertainty in the space-time distribution of rainfall had a much more profound effect on runoff predictions than uncertainty in watershed characteristics. The most sensitive land-surface parameters were the soil-saturated hydraulic conductivity, the fraction of impervious area, and the retention depth (i.e., the capacity of the land surface to hold water in microtopographic features).

Hydraulic Processes

Rainwater that is not absorbed by the land surface starts running off hill slopes (at slow flow velocity) and eventually reaches a channel (attaining fast flow velocity), where it gets routed through the drainage network to the catchment outlet. The role of the network structure in controlling a basin-scale flood response can be evaluated through the geomorphological instantaneous unit hydrograph (GIUH; Rodriguez-Iturbe and Rinaldo, 1997), which represents the basin response at time *t* to a unit mass input of rainfall at time $t = 0$ uniformly distributed over the catchment. The velocity of water flowing through a channel plays a fundamental role in determining the timing of flood response to rainfall input. A basin's response time to rainfall input may be estimated by the runoff volume-to-peak discharge ratio (Potter, 1991; Bradley and Potter, 1992; J. A. Smith et al., 2000), which can be calculated from historic stream gauging records of annual peak discharge and mean daily discharge values. The lag time—defined as the time difference between the time centroid of rainfall and peak discharge—provides a useful time scale for the analysis of space-time variability of rainfall over a catchment. The lag time for a drainage basin can be viewed as an upper bound on the time scale of rainfall distribution that is relevant to flood magnitudes at the basin outlet (J. A. Smith et al., 2001).

For warning purposes, the prediction of peak discharge, especially the time to peak, and of discharge volume is crucial. The peak discharge reveals how much the water level may rise above riverbanks or man-made drainage system confines, while the time from the onset of rising water levels to its peak indicates how much time might be available for evacuation and other protective actions. In addition, the discharge volume provides a measure of how long it will take for the water level to drop off to a normal stage again.

The temporal and spatial organization of rainfall plays a fundamental role in flash flood hydrologic behavior. The rainfall properties that are important for a given drainage basin depend on the network properties of the drainage basin and the velocity of the water within the channel. Storm structure and motion interact with drainage network geometry to control flash flood response (J. A. Smith et al., 1996, 2000). An important factor determining flood response is the spatial and temporal extent to which a basin is covered by heavy rainfall. A large fractional coverage of heavy rainfall for durations corresponding to the basin response time results in maximum peak discharges (J. A. Smith et al., 2002). On the other hand, storm motion relative to a catchment may either amplify or dampen a flood wave. Movement of a storm down the basin results in anomalously large peak discharges for a given rainfall and runoff. For example, another factor in the aforementioned Buffalo Creek, Colorado, flash flood was that the focusing and timing of the rainfall and storm movement coincided with the movement of the flood wave along the valley (Landel et al., 1999). In essence, storm movement acted in concert with flood wave propagation to amplify peak discharges.

If the peak discharge exceeds the riverbank's holding capacity, then water will spread out into the floodplain. Floodplain inundation is a major environmental hazard that is not well understood and lacks consensus with regard to the level of model and data complexity required to achieve a useful prediction of inundation extent, water surface elevation, and sediment deposition (Hughes, 1980; Nicholas and Walling, 1997; Horritt and Bates, 2001; Romanowicz and Beven, 2003; Jothityangkoon and Sivapalan, 2003). When the peak discharge exceeds river capacity and the floodplain is inundated, a channel-dominated flow transitions into a valley bottom-dominated flow, which tends to attenuate the flood wave propagation and thus increase lag time, depending on floodplain topography and surface roughness. Flood peak attenuation is largely a result of storage (or greatly reduced velocity) of a portion of the runoff on overbank surfaces (Woltemade and Potter, 1994). This storage and the later release of a portion of the total flood volume produce flood hydrographs that are low and broad compared to those of similar watersheds that lack floodplain storage, such as gullies or mountain streams. Attenuation can play a major role in regional flood hydrologic behavior (Shiono et al., 1999; Zhang et al., 2001; Turner-Gillespie et al., 2003). Natural attenuation of flood waves results primarily from geologically controlled variations in the longitudinal profile of the river channel network and the valley bottom width (Wolff and Burges, 1994; Woltemade and Potter, 1994; J. A. Smith et al., 2002; Turner-Gillespie et al., 2003). Typically, flood wave attenuation increases with increasing valley bottom width and decreasing stream slope. Detailed analyses by Woltemade and Potter (1994) reveal that moderate-magnitude floods (5- to 50-year recurrence interval) with relatively high peak-to-volume ratios are attenuated most, since the storage of a relatively small volume of water can significantly reduce the peak discharge. In contrast, both small and large floods are attenuated relatively little.

Impact of Urbanization

Urbanization significantly affects hydrometeorological, hydrological, and hydraulic processes. For example, pollution and urban heat island effects may alter the dynamics, thermodynamics, and microphysics of precipitation formation; rapidly expanding impermeable surfaces affect hydrologic processes at the land surface; and stormwater management and drainage network construction modify the hydraulics of water flowing in natural and man-made channels. Urban flooding is becoming an increasingly serious problem because of the removal of vegetation, placement of debris in channels, construction of culverts and bridges that constrict flood flows, paving and other replacement of ground cover by impermeable surfaces that increase runoff, and construction of drainage systems that accelerate runoff (Gruntfest and Huber, 1991).

The flood response for a catchment can be altered significantly through urbanization, as demonstrated by J. A. Smith et al. (2002) for the Charlotte, North Carolina, metropolitan area. The hydrologic response to urbanization typically is characterized by increasing flood peak magnitudes, decreasing lag time, and increasing runoff volumes (Leopold, 1968; J. A. Smith et al., 2002). The timing and magnitude of flood peaks can be very sensitive to alteration of the drainage network, which increases the drainage density of the basin and the hydraulic efficiency of the drainage system (Graf, 1977; Hollis, 1988). Stormwater control structures, such as detention basins, have a significant impact on the flood discharge downstream. They typically are designed to shave off the peak of extreme discharges to keep the water within the natural or man-made channel network confines, but they may result in high-water-stage levels for a prolonged period of time.

Urban sprawl results in rapidly changing hydrologic and hydraulic conditions, making it difficult to assess the relative timing of local and upstream contributions to flood response and, consequently, to determine the cumulative flood response of a basin. Moreover, overflowing sewer and drainage systems and street flooding increasingly threaten the population living in urban areas. Issuing timely warnings for flash floods, therefore, remains a challenging task more so than ever before.

3

The National Weather Service and Flash Floods

The National Oceanic and Atmospheric Administration's (NOAA's) National Weather Service (NWS) is the federal organization charged with protecting the nation from extreme weather events, including flash floods. Specifically, the mission of the NWS is to provide ". . . weather, water, and climate forecasts and warnings for the United States, its territories, adjacent waters and ocean areas, for the protection of life and property and the enhancement of the national economy" (NWS, 2003a). There are 122 NWS Weather Forecast Offices (WFOs)—116 in the continental United States, 3 in Alaska, 1 in Hawaii, 1 in Puerto Rico, and 1 in Guam (Figure 3.1)—divided into six regions. Each WFO serves a designated group of counties, ranging from a few to more than 20, termed the county warning area (CWA). For example, the Salt Lake City WFO serves 25 counties, whereas the Los Angeles-Oxnard WFO is responsible for only 4 counties; the Los Angeles-Oxnard WFO, however, serves a much greater population than Salt Lake City's WFO. The WFOs are responsible for providing flash flood forecasts, watches, and warnings for their specified CWAs. In addition to the WFOs, there are 13 River Forecast Centers (RFCs) and 9 national specialized centers that provide support to NWS functions. In particular, the Hydrometeorological Prediction Center (HPC) is especially relevant for supporting the NWS prior to and during heavy precipitation events. This chapter gives additional information about the NWS and its role pertaining to flash floods, including a brief description of the modernization effort that led to the development of the NEXRAD network and of the extensive process that local NWS meteorologists go through to forecast and warn of flash flood events.

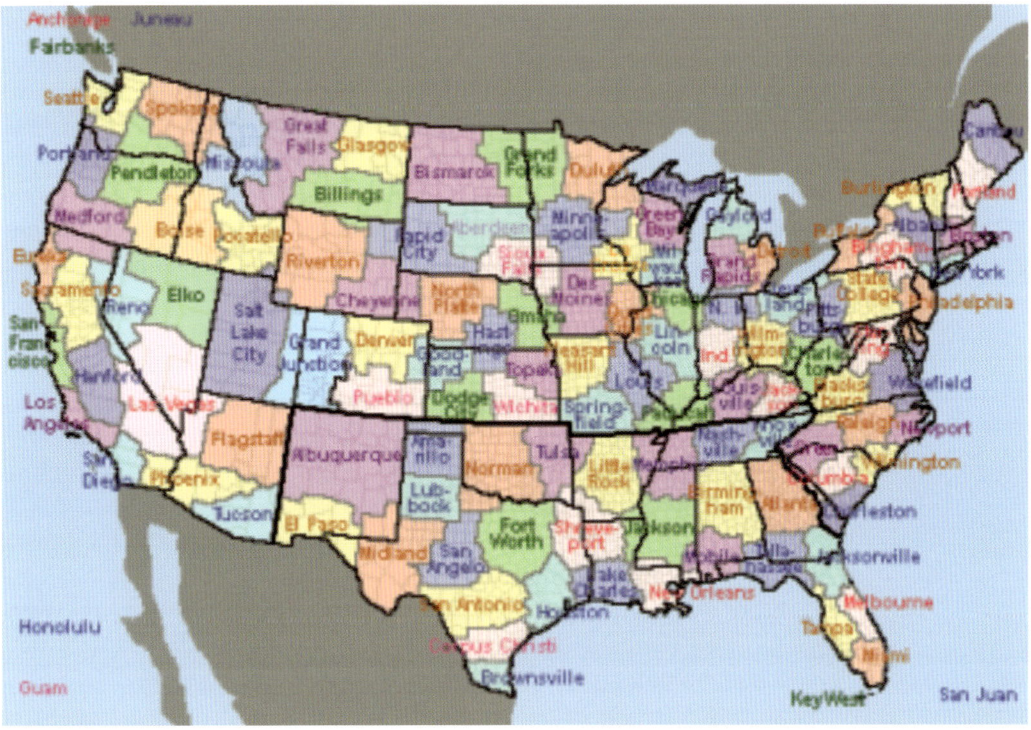

FIGURE 3.1 Distribution of the 122 National Weather Service WFOs in the continental United States, Alaska, Hawaii, Guam, and Puerto Rico. The CWAs that each WFO serves are shaded. The Southern California coastal area shaded in pink is the CWA served by the Los Angeles-Oxnard WFO. SOURCE: National Weather Service.

THE MODERNIZATION AND ASSOCIATED RESTRUCTURING OF THE NATIONAL WEATHER SERVICE

Vast improvements in instrumentation, computer, communications, and data processing technologies since the 1950s led to much improved weather observing technologies in the 1970s and 1980s. These advances, together with a better scientific understanding of atmospheric processes, allowed for major improvements in NWS capabilities to accurately observe, forecast, and warn of severe weather. To capitalize on the scientific and technological advances for improving its services, the NWS underwent a major modernization and associated restructuring (MAR) during the 1990s. This modernization was the most important development in operational meteorology in the past 50 years (for details of the MAR, see Friday, 1994). A major outcome of the MAR was that it improved local data and guidance analyses available to WFOs. Because local forecasters know best the local topography and weather conditions that produce hazardous weather in their regions, the improved data and guidance allowed for increased accuracy and timeliness of their service to the public and reduced losses of lives and property. Each WFO has four major responsibilities, including:

1. issuing forecasts, watches, and warnings for severe local weather hazards, flash floods, floods, heat waves, winter storms, fire weather conditions, agricultural weather conditions, and all other local weather information;
2. issuing local aviation weather forecasts and warnings as well as forecasts for en route pilots in the United States;
3. issuing marine forecasts, watches, and warnings for all U.S. coastal areas and the Great Lakes; and
4. providing hydrologic services, including issuing watches and warnings for flash floods and slow-rising floods, identifying flash flood-prone areas, and monitoring local hydrological surveillance systems.

The modernization also led to a change in staffing nationwide by creating a core of professional meteorologists and hydrologists at the WFOs and RFCs to most effectively utilize the new data. Each WFO has a Science and Operations Officer, and the RFC has a corresponding Development and Operations Hydrologist. Their jobs are to oversee the expedient transfer of emerging scientific and technical discoveries from researchers to daily NWS and WFO operations, by collaborating with local universities and research laboratories. Each WFO also has a Warning and Coordination Meteorologist who maintains close contacts with local emergency managers to effectively disseminate watches and warnings to local government officials responsible for alerting the public and other users of all existing and potential weather-related hazards. A Service Hydrologist at each WFO manages the hydrological warning program and coordinates closely with the regional RFC and the HPC for hydrological guidance.

The MAR focused on many major new technologies to exploit the improved scientific and technological capabilities. In the computing arena, National Centers for Environmental Prediction (NCEP) supercomputers were slated to perform four-dimensional data assimilation techniques and to run numerical weather prediction (NWP) models that provide forecasts and guidance for periods beyond 36 hours. A new, high-speed computer workstation, termed the Advanced Weather Interactive Processing Systems (AWIPS), was developed and now serves as the nerve center of operations at all WFOs and RFCs. AWIPS receives all observational data and guidance analyses from the national centers and provides an array of current and forecast information for use by local WFO forecasters to assess the need for forecasts and warnings. Once the forecaster makes a decision, AWIPS formulates and disseminates warnings and forecasts automatically through a communications system, making them available at all NWS sites and to emergency managers, private weather firms, and universities.

Expanding the spatial network of atmospheric observations was an important objective of the MAR, and because of the significant effects of weather on aviation and defense, these efforts were supported greatly by the Federal Aviation Administration (FAA) and the Department of Defense (DoD). To accomplish this, the Automated Surface Observing System and wind profilers were developed, instrumentation on geostationary and polar-orbiting satellites was improved, and most relevant to this study, the Weather Surveillance Radar-1988 Doppler (WSR-88D) radar called the Next Generation Radar (NEXRAD) was deployed. NEXRAD is the pioneering technology used for detecting and tracking of severe mesoscale storms, such as tornadoes and severe thunderstorms, that often lead to flash floods. It can detect mesocyclones within clouds aloft, identifying a precursor to subsequent tornadic activity and thereby providing longer tornado warnings. NEXRAD can also track more widespread precipitation and give precipitation estimates, increasing the accuracy and timeliness of flash flood forecasts. Chapter 4 provides details of the NEXRAD network and its utility.

THE PROCESS OF ISSUING FLASH FLOOD FORECASTS, WATCHES, AND WARNINGS

The NWS serves as the sole official voice for issuing warnings during life-threatening weather situations. The flash flood forecasts, watches, and warnings are issued only by the WFOs, but the RFCs and the national centers—such as the NCEP, the HPC, the Storm Prediction Center, and the National Hurricane Center—provide essential supporting information and guidance to the WFOs. Each warning is issued for a specific time period and for a specific CWA, although a given warning statement may cover more than one CWA. The warning statements, however, often identify specific threatened subareas within a large CWA, such as those in the Los Angeles area. Additional flash flood statements may provide supplementary updates about ongoing events, and it is common to have a series of extensions to an initial warning if a rainy period covers many hours.

The NWS process for issuing flash flood forecasts, watches, and warnings has changed markedly in the last 10 to 15 years, largely as a consequence of the NWS modernization. In the past, forecasters simply analyzed past weather and hydrological data to determine how much precipitation fell in a given area (e.g., to assess current soil moisture conditions) as well as what happened to that rainfall (e.g., to see if flash floods recently occurred). Then the forecaster would compare that information with predictions of future rainfall and attempt to subjectively assess flash flood potential. Although the general idea remains the same, the collection of more and newer remotely sensed and in situ data, combined with continually improving probabilistic and deterministic tools, has greatly improved NWS procedures and lead times for issuing flash flood forecasts, watches, and warnings during the past 15 years.

Successful forecasts of a heavy rainfall event depend on interpretation of the meteorological and hydrological aspects of a particular situation. Forecasters must first determine how much precipitation will fall, where, when, and of what intensity. Then this information is paired with knowledge of the antecedent moisture conditions of the soil and the stages of streams and rivers in the forecast area to gauge the likelihood of flooding.

Output data and forecast fields from NWP models are key tools that allow the forecasting process to begin several days before an event occurs. A variety of NWP models are run on an operational basis by the NWS, and their output is used as guidance by weather forecasters. The models have somewhat different numerical formulations of the various atmospheric processes and, hence, sometimes arrive at significantly different forecasts. One of these numerical models—the Global Forecast System model—is run more than once, and sometimes several times, per set of input data in what is called an ensemble approach. The ensemble approach allows small modifications to the input data or other small intentional changes to the initial conditions to test the sensitivity of the forecasts to small changes in initial conditions or model formulations. Statistical packages are used to postprocess NWP model forecast data to eliminate potential numerical model biases, adjust for local factors, and output additional quantities not directly predicted by the numerical models. Other numerical models are run operationally at least daily by the U.S. Navy and U.S. Air Force, by foreign government weather services, by local NWS weather forecast offices, and by university research programs. All of this output can be examined by local NWS forecasters. Forecasters then assess the predicted values of numerous weather variables and their model-to-model variations and determine the most probable upcoming weather. Forecasters will usually draw on their experience of how the numerical models handle different weather situations, along with their experience of how various weather situations tend to evolve, especially in their local forecast regions, given the initial weather pattern.

A multitude of types of data, collected around the world, provide input to the NWP models. Measurements are made of air pressure, temperature, humidity, wind direction and speed, cloud coverage, and cloud base heights from observations at surface weather stations, buoys, and ships. Some of these measurements are available several times per hour. Information about conditions more than a few meters above the ground is obtained from a variety of platforms. Twice per day, nearly 100 rawinsondes are launched on weather balloons with instrument packages that transmit temperature, humidity, pressure, and wind data at numerous altitudes back to the NWS office as they ascend. As part of the Aircraft Communication Addressing and Reporting Systems, some commercial aircraft are instrumented to transmit wind, temperature, and humidity data while they ascend and descend near airports and during their cross-country transits. Vertically pointing radar wind profilers at more than three dozen locations nationwide measure winds hourly up to 16-km altitude and down to 500 m above the ground (i.e., for 400-MHz band profilers) or, when combined with Doppler sodars, down to 30 m (i.e., for 900-MHz band profilers). When combined with radio acoustic sounding systems, wind profilers also provide vertical temperature profiles at hourly intervals. Doppler weather radar data also can be used to determine the average vertical profile of winds above the radar sites in many situations. Many types of data are available from geostationary and polar-orbiting satellites that serve as the best sources of data over the oceans in terms of spatial and temporal resolution. Often, satellite information is used to assess the accuracy of analyses used in the numerical model initialization process and to obtain a rough assessment of the character of the approaching weather system. Visible and infrared satellite imagery can be used to determine the locations and altitudes of clouds. Water vapor imagery is used to measure the middle- and upper-level water vapor content of the atmosphere. Microwave data are used to determine the water content of the cloud and precipitation systems, especially over oceans, and satellite sounder data, using radiation measurements at a host of wavelengths, are used to obtain temperature profiles.

All of these data then are assimilated into the numerical models, in a procedure that simultaneously invokes a measure of quality control. New observed data values are examined against values from a previous numerical model forecast—a "first-guess" value for the observation time. If a new measurement is within reasonable agreement with the first-guess value, it is used to modify the first-guess value in the data assimilation process. Most numerical models make computations at dozens of levels in the vertical and at grid points spaced about 40 km apart in the horizontal (varying among NWP models). Forecasters use precipitation forecasts from the numerical models as guidance in making a forecast of the timing and amount of upcoming precipitation. Although the skill of the models in forecasting heavy rainfall in the warm season is extremely low (Fritsch and Carbone, 2004), the models often give reasonable portrayals of upcoming storms in the cool season. In any case, human forecasters typically adjust and improve on the numerical guidance by using their experience, including dealing with the influence of regional terrain on the upcoming weather system.

When it is determined that rainfall is imminent, forecasters must assess flash flood potential in order to decide whether and when to issue flood and flash flood watches and warnings. The regional RFC and the HPC are important sources for assessing the hydrologic situation, with their combined knowledge of the local topography and soil and stream conditions. They provide daily flash flood guidance to the WFOs on hydrologic conditions, including information such as localized assessments of the amount of rainfall that will lead to flooding in a given area. This guidance, combined with anticipated and current rainfall rates, helps WFOs to best gauge local flood potential.

Use of a decision matrix is another way to assess flash flood potential. For example, inputs for convective storms might include such parameters as precipitable water amounts in various categories, the 500-mb steering-level wind speed, the existence of a cap on a stable layer aloft (i.e., at approximately 500 mb), and the existence of a trigger mechanism such as a front, outflow boundary, or topography. Table 3.1 shows an example of a flash flood potential decision matrix. For nonconvective storms, other weather variables also are important, but precipitation amounts and intensities still are key parameters. The speed of the upper-level winds normal to mountain barriers is an important determinant of a lifting mechanism that often produces rain in stratiform clouds near mountains. Sometimes convective elements are embedded in stratiform clouds, so both the convective and the stratiform weather parameters must be examined.

Because rainfall patterns often move across an area rather than develop locally, measurements from Doppler radars such as NEXRAD generally are the most valuable information for monitoring approaching rainstorms. Sat-

TABLE 3.1 Decision Matrix to Assess Flash Flood Potential

500-mb Wind (knots)	Cap	Trigger	PW > 1.00	0.80 < PW < 1.00	0.50 < PW < 0.80	PW < 0.50
<15	No	Yes	Very high	High	Moderate to high	Low
		No	High	Moderate to high	Moderate	Low
	Yes	Yes	Moderate to high	Low to moderate	Low	Low
		No	Low to moderate	Low	Low	Low
15–25	No	Yes	High	Moderate to high	Moderate	Low
		No	Moderate to high	Moderate	Low to moderate	Low
	Yes	Yes	Moderate to high	Low to moderate	Low	Low
		No	Low to moderate	Low	Low	Low
≥ 25	No	Yes	Moderate to high	Moderate	Low to moderate	Low
		No	Moderate	Low to moderate	Low	Low
	Yes	Yes	Low to moderate	Low	Low	Low
		No	Low	Low	Low	Low

SOURCE: Brian McInerney, NWS Salt Lake City WFO.

ellite data also can be used, but the temporal and spatial resolutions are not as high and precipitation estimates are less reliable. Networks of rain gauges at the surface are used to measure how much rain has fallen, but the gauges generally are spaced dozens of miles or more apart. Small-scale heavy rain cores and bands that often lead to flash flooding can occur between gauge locations and, thus, result in undersampling of the actual rainfall. Moreover, only a subset of gauges provides real-time reporting of the observations. For these reasons, forecasters rely heavily on radar estimates of current and approaching rainfall intensities and accumulated amounts—ground-truth-checked against available rain gauge data when and where possible—in the warning process, particularly to identify small-scale areas of maximum precipitation. However, radar measurements of rainfall rate and accumulation must be considered only estimates. The radar actually measures the power returned to its receiver, as reflected back by whatever targets were in the path of the beam, and uses this information to infer precipitation rate by assuming an idealized (i.e., typical) relationship to the distribution of number and sizes of raindrops within the beam.

In situ rainfall measurements are acquired by various types of rain gauges. Some are buckets (i.e., standard 8-inch rain gauges) that require a human to insert a ruler-like instrument to measure the depth of the accumulated rainfall. Others include tipping buckets and weighing gauges, which may include a moving chart that records the amount of precipitation that has fallen as a function of time. The tipping bucket is so-called because a small bucket on an arm-like device tips and empties each time 0.01 inch of rain accumulates. The weighing and tipping-bucket types of rain gauges can be equipped to report automatically to the NWS each hour or more frequently, whereas many gauges staffed by human observers may report only once per day unless special provisions are made for additional reporting in heavy rain situations.

Although it may seem that rain gauges should provide exact rainfall amounts, in practice their values can be in error or unrepresentative. During high rainfall rates, the tipping bucket may not be able to keep up with the intensity of the rain and result in an underestimate. In high wind conditions, some rain may blow over the gauge, even if wind shields are positioned around it. Spatial variations in rainfall due to topographic or other factors may also make the gauge value unrepresentative of the basin-average rainfall that contributes to flood conditions. Thus, the blending of radar and rain gauge data tends to give the most complete picture of the basin-average rainfall.

During a rain event, operational meteorologists and hydrologists also monitor measurements of streamflow and stage that are available at strategic locations on streams and rivers. Streamflow indicates the volume of water flowing down the channel in a fixed amount of time, such as cubic meters per second. Stage indicates the water level in the stream or river and is compared against flood stage values, beyond which the water overflows the banks and flooding begins. In the case of slower-evolving stream and river floods, as opposed to flash floods, NWS forecasters have used floodplain mapping and experience to determine what areas will be affected for various stage values, and they are able to convey this information to the public in flood warning messages. Reports also are received from trained spotters and observers, emergency managers, law enforcement officials, and the public to indicate where flooding has begun or is imminent.

In summary, NWS forecasters go through an extensive, multiday process of assessing the flash flood potential of approaching weather systems. They monitor numerous data from a myriad of sources and assess models and real-time observations in order to determine whether a flash flood situation is developing. The forecast process is highly regionalized in that forecasters' assessments are based on their local knowledge and experience, which often serve as the most valuable and effective forecasting tools.

4

NEXRAD

THE NEXRAD NETWORK[1]

Appendix B summarizes the technical characteristics of the NEXRAD system. Crum and Alberty (1993) and Serafin and Wilson (2000) provide additional background on the system characteristics. Coverage over the eastern two-thirds of the country is essentially complete, although significant limitations exist in coverage near the surface (NRC, 1995). The NEXRAD coverage sometimes is portrayed in terms of the coverage at 3.05 km (10,000 ft) above the level of the radar sites. The 3.05-km (10,000-ft) criterion was based on language in the Weather Service Modernization Act (P.L. 102-567; Sec. 702(4)), and the above-site-level extension appears to have been an interpretation of that language. The extent of near-surface coverage was a consideration in the NEXRAD site selection process (Leone et al., 1989), but the 3.05-km (10,000-ft) above-site-level representation does not convey a true picture of the low-level coverage. There is considerable variation in above ground level (AGL) or above mean sea level (MSL) coverage where sites were established in mountainous terrain. Thus, there are some substantial gaps in western regions (see, for example, Figure 4.4), and the combination of high-altitude sites and mountainous terrain presents difficult problems in several areas (Westrick et al., 1999; Maddox et al., 2002).

Primary Data and Derived Products

The NEXRAD is a pulse-Doppler system that measures three primary characteristics of the radar echoes: the radar reflectivity factor, commonly

[1]Adapted from NRC (2002).

referred to as reflectivity and designated by *Z*; the Doppler (radial) velocity, designated by *v* or v_r; and the width of the Doppler spectrum, designated by σ_v. These base data variables, derived in the radar data acquisition (RDA) unit, express the zeroth, first, and second moments, respectively, of the Doppler spectrum of the echoes. A value for each quantity is available for every resolution cell of the radar, as defined by its antenna beamwidth and the sampling rate along the beam axis, although the latter is constrained presently to no less than half the pulse duration.

Displays of these quantities, together with other products (Radar Operations Center, 2002) and results of the algorithms discussed below, are developed from the base data in a radar product generator (RPG) unit. The products include estimates of precipitation accumulations for 1-hour and 3-hour periods as well as a storm-total product. In addition, a series of computer algorithms operates on the base data—and some also incorporate auxiliary information such as temperature profiles—examining the echo patterns and their continuity in space and time in order to identify significant weather features such as mesoscylones, tornado vortex signatures, or the presence of hail. Outputs of these algorithms are displayed as icons superimposed on the basic radar displays or in auxiliary tables. The number and the variety of potential algorithms continue to increase as scientific knowledge about the relationship between echo characteristics and storm properties improves and available computational resources increase.

Data Display, Dissemination, and Archiving

A principal user processor (PUP) associated with each NEXRAD installation, and numerous additional remote PUPs, provided the initial data display capability. The PUP was essentially a mainframe minicomputer, with a monitor, that operated programs to generate displays from a rather limited set of possibilities. As computer technology has advanced, open-systems architecture has been implemented to replace both the RPG and the PUP units. Thus, the current open RPG (ORPG) generates and displays the various products as well as relays the relevant data for display on other systems.

Although the NEXRAD system of radars is of major value as a stand-alone weather-observing network, additional value is obtained through the integration of NEXRAD data with other weather observations (e.g., wind profilers, satellites, the National Lightning Detection Network, surface measurements, other radar systems) and associated analyses. As described earlier, this synthesis is carried out in the National Weather Service (NWS), the Advanced Weather Interactive Processing System (AWIPS), and dissemination of NEXRAD data within the NWS is handled through the AWIPS system. Equivalent systems support the Federal Aviation Administration (FAA) and Department of Defense (DoD) users of NEXRAD data. Dissemination to outside users, formerly handled by vendors, is now accomplished through the Base Data Distribution System (BDDS), with distribution over the Internet. AWIPS and similar systems are expected to mature dramatically and grow in use in the future as the science of meteorology and the technology of information processing, display, and dissemination continue to advance and merge with societal needs for improved weather information and forecasting.

Archiving of the NEXRAD base data—referred to as Level II data—was previously accomplished by magnetic tape recording at the sites, with the tapes being shipped to the National Climatic Data Center (NCDC) (Crum et al., 1993). Experience showed that only a little more than half of the national dataset reached the archive in retrievable form. Results of the Collaborative Radar Acquisition Field Test (CRAFT; Droegemeier et al., 2002) established the capability to transmit NEXRAD data over the Internet to NCDC and increase the fraction of retrievable data. A separate archive of a set of the derived products—referred to as Level III data—provides basic data for such things as research, training, and legal inquiries.

Users and Uses of the Data and Products

The principal NEXRAD user agencies are the Department of Commerce (DOC), DoD, and Department of Transportation (DOT). The primary mission organizations within these agencies are the NWS, the Air Force Weather Agency (AFWA), the Naval Meteorology and Oceanography Command (NMOC), and the FAA. As discussed in Chapter 3, the NWS is responsible for the detection of hazardous weather and for warning the public about these hazards in a timely, accurate, and effective way. The NWS also provides essential weather information in support of the nation's river and flood prediction program, as well as in support of civilian aviation, agriculture, forestry, and marine operations. The national information database and infrastructure formed by NWS data and products can be used by other government agencies, the private sector, the public, and the global community. The AFWA provides worldwide meteorological and airspace environmental services to the Air Force, Army, certain other DoD organizations, and intelligence agencies. NMOC supports the U.S. Navy, U.S. Marine Corps, and certain other DoD organizations. The primary missions of these

DoD agencies are to provide timely information on severe weather for the protection of DoD personnel and property; to provide weather-related information in support of decision-making processes at specific locations; and to support military aviation. The FAA's responsibility requires it to gather information on the location, intensity, and development of hazardous weather conditions as well as to provide this information to pilots and air traffic controllers and managers.

The group of users has expanded dramatically to include, among others, a very large atmospheric sciences and hydrometeorological research community in universities and research laboratories throughout the world; other federal, state, and local government organizations and private sector providers; and distributors and users of weather and climate information gleaned from meteorological radar measurements and associated products (see NRC, 2003). The latter include data that are either taken or derived directly from radar measurements, as well as information derived through intelligent integration of radar data with other measurements and analyses of weather events.

Limitations of the System

A variety of limitations impede the ability of the NEXRAD system to meet the needs of all of its varied users. Some limitations, such as the divergence of the radar beam with increasing range, are inherent to any radar system. Others, such as the inability to acquire data in small elevation steps during shallow winter precipitation episodes, can be overcome by rather straightforward hardware or software modifications, the latter of which will be facilitated by the greater flexibility afforded by the open-systems architecture. The ongoing program of research and development should provide at least partial solutions to some of the other problems.

Serafin and Wilson (2000) provide a good summary of the recognized deficiencies of the NEXRAD system. Those that affect the primary variables directly include contamination by ground clutter, both that in the normal radar environment and that arising during anomalous propagation (AP) conditions, and the occasional impact of bird echoes on the Doppler velocity data. The problem of range-velocity folding, common to all pulse-Doppler radars, has proved to be quite serious in the NEXRAD system. Studies of several techniques now under way should yield means of mitigating the range-velocity folding problem.

Spatial coverage limitations are imposed in the first instance by the curvature of Earth. This limitation constrains the available coverage to minimum altitudes that increase with distance from the radar site. The problem is exacerbated by any obstacles in the radar environment that constitute a radar horizon extending above 0.0° elevation angle. A further constraint currently imposed on the NEXRAD system limits the minimum elevation angle to no lower than 0.5°, adding to this difficulty. This problem is of special concern for radars at high-altitude sites in mountainous areas, such as the Sulphur Mountain NEXRAD. Similar difficulties arise in areas subject to intense precipitation from shallow cloud systems, such as places in the lee of the Great Lakes affected by lake-effect snowstorms. NEXRAD scans also are restricted to some maximum elevation angle (currently 20.0°), mainly to provide an acceptable scan update rate (see below), but the result is a "cone of silence" data gap above each radar site.

Regions devoid of data pose a difficult problem for NEXRAD algorithms. The primary reasons for data voids are beam overshoot, beam blockage due to obstructions, the cone of silence near the radar, and gaps in vertical coverage arising when large elevation steps are used between azimuth traverses in the scan strategy, along with regions of low echo strength, data masking due to data corruption, and planned and unplanned outages. Data voids resulting from overshoot, beam blockage, the cone of silence, and vertical gaps are determined by the geometry of the radar installation and the scan strategy. Overshoot leads to a data void caused by the elevation of the lowest beam above the surface. Beam blockage and the cone of silence prevent the acquisition of data from affected regions of the atmosphere. Gaps in vertical coverage result from the usual scan strategies that have fairly coarse vertical beam spacing at high elevation angles to achieve more rapid volume scans. The trade-off here is between accepting longer times for the volume scan, accepting larger vertical gaps (i.e., fewer tilts), or enlarging the cone of silence by limiting the elevation of the highest tilt. Some products are more tolerant of vertical gaps than others. Early termination of volume scans by NEXRAD operators seeking more rapid updates of low-level base data occasionally introduces additional voids in the high-level data.

In regions of low echo strength due to the scarcity of reflectors, the signal can become so weak that the wind velocity cannot be resolved. The resulting velocity data void can affect the products involving the velocity data. Some compensation is possible by changing the waveform or the scan strategy, as in the present clear-air mode volume coverage pattern (see the following section for a discussion on volume coverage patterns).

A final deficiency concerns NEXRAD precipitation estimates (J. A. Smith et al., 1996; Anagnostou et al., 1998), which are important to a variety of applications. There are fundamental problems in converting the measured reflectivity to precipitation rates, as discussed in depth later in this chapter.

NEXRAD SCAN STRATEGIES

Surveillance of the atmospheric volume surrounding a NEXRAD site is provided through one of several available scan routines known as volume coverage patterns (VCPs). The VCPs summarized in Table 4.1 are commonly used. The clear-air patterns cover the lowest layers of the atmosphere in 10 minutes and provide such things as wind profiles and indications of sea breeze fronts or storm outflow boundaries that could trigger convective activity. The "precipitation" and "severe weather" patterns cover the full depth of storm activity in 5 to 6 minutes and provide more frequent updates on evolving storms. Although these update cycles may be limiting during rapidly evolving convective weather, such update rates are adequate to capture relevant precipitation features for flash flooding (e.g., J. A. Smith et al., 1996, 2000).

The elevation steps between scans at the low end of the VCPs are usually 1°, or a bit less (i.e., essentially equivalent to the radar beamwidth, which usually is defined as the angular distance between points on either side of the beam axis where the power density is half the value along the axis). This corresponds roughly to the Rayleigh resolution criterion[2] for distinguishing a pair of adjacent point sources (or targets, in the radar domain). However, it does not provide the most complete depiction of storm structures that are distributed in the vertical. As noted by Brown et al. (2002), smaller elevation increments at the lowest angles would provide additional useful information, especially for echoes at long range that may extend only up into the first few elevation steps of the current VCPs. The forthcoming open RDA-RPG combination is capable of implementing a more flexible variety of VCPs, which will include ones with minimum elevation steps of order 0.5° as suggested by Brown et al. to accommodate this concern. One such VCP is to be implemented in the NEXRAD network in 2004. The smaller elevation steps will be useful in estimating precipitation in many instances, such as when the beam is blocked at the lowest tilt angle.

---

[2]The Rayleigh resolution criterion for the resolving power of an antenna is that two points can just be resolved (i.e., distinguished as two points rather than one) when their angular separation, θ, is given by $\theta = 1.22*\lambda/D$, where λ is the wavelength of the radiation and D is the diameter of the receiving antenna (Kidder and Vonder Haar, 1995).

TABLE 4.1 NEXRAD Volume Coverage Patterns[a]

Scan Strategy	Number of 360° Azimuthal Scans	Number of Unique Elevation Steps	Elevation Range of Azimuthal Scans	Time to Complete (min)
Clear air (short pulse)	7	5	0.5° to 4.5°	10
Clear air (long pulse)	8	5	0.5° to 4.5°	10
Precipitation	11	9	0.5° to 19.5°	6
Severe weather	16	14	0.5° to 19.5°	5

[a]The azimuthal scan at the two lowest elevation angles (three for clear air long pulse) is repeated to permit one scan in a low-PRF (pulse-repetition frequency) surveillance mode (to map the reflectivity field) and another in a high-PRF Doppler mode (to measure radial velocities). At higher elevations, these functions are done during the same azimuthal scan. SOURCE: Adapted from Crum et al., 1993.

Elevation steps greater than the radar beamwidth at the high elevation angles in those VCPs, plus the cone of silence over the radar site itself, lead to some data voids, particularly for echoes near the radar. Although the low-level coverage permits estimates of precipitation at the height of the beam, such data voids can interfere with attempts to use the vertical profile of reflectivity as an aid in projecting those estimates down to the surface.

Problems with low-level coverage from high-altitude radar sites in mountainous terrain were noted in the National Research Council NEXRAD coverage report (NRC, 1995) and have been discussed further by Westrick et al. (1999), NRC (2002), and Maddox et al. (2002). There is no easy compromise to achieve both low-level and wide-area radar coverage in mountainous terrain, but the current constraint to a minimum elevation angle of 0.5° clearly exacerbates the problem. It has already been noted that the use of lower, even negative, minimum elevation angles would mitigate the problem for NEXRAD sites in Montana (Brown et al., 2002) and Utah (Wood et al., 2003b). The utility of this approach in other situations depends on the degree of beam blocking that might result in the directions of importance. The NWS has abandoned the use of the hybrid scan rainfall estimation scheme (Fulton et al. 1998) in western mountainous areas. Only the lowest nonblocked elevation angle is now used. Thus, a modification of the scan scheme to use negative elevation angles may not pose a huge algorithmic problem.

PRECIPITATION ESTIMATION

Estimation of rainfall using radar data has been a well-researched topic (Box 4.1) and has received extensive attention in the literature. The main

BOX 4.1

Radar Basis for Characterizing Precipitation

The scattering of electromagnetic waves by precipitation particles and their propagation through precipitation media form the basis of radar-based characterization of precipitation. Figure 4.1 below shows the geometry of the scattering volume within the precipitation medium for ground radars. The power received at the radar from precipitation is composed of the backscattered power contribution from all of the particles in the radar resolution volume. Therefore, it is useful to work with radar cross sections per unit volume that can be related to microphysical properties of precipitation. At the S-band frequencies (i.e., 2.7–3.0 GHz) at which NEXRAD operates, the volumetric radar cross section, η, is given by

$$\eta = \frac{\pi^5}{\lambda^4}\left|K_p\right|^2 \int N(D) D^6 \, dD, \tag{1}$$

where the dimensionless factor $\left|K_p\right|^2 = \left|\frac{\varepsilon_r - 1}{\varepsilon_r + 2}\right|^2$; ε_r is the complex dielectric constant of the precipitation particle, and (1) assumes that all of the particles are composed of the same material (e.g., water). *N(D)* is the particle size distribution and *D* is the equivolume diameter of a particle.

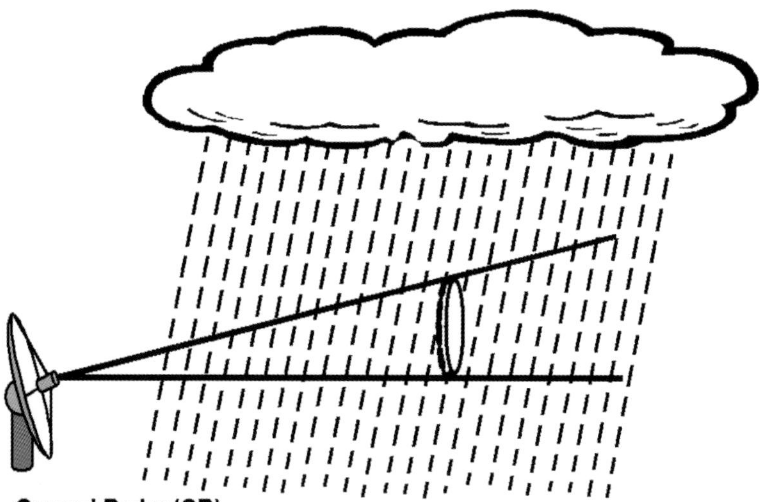

FIGURE 4.1 Geometry of ground-based radar, where the shaded area shows the resolution volume. SOURCE: V. Chandrasekar, Colorado State University.

The reflectivity factor Z is defined as

$$Z = \int_D N(D) D^6 \, dD \tag{2}$$

The received power is determined by the backscattering properties of the hydrometeors in the resolution volume as well as by the forward-scattering properties of the particles over the propagation path from the radar to the resolution volume. From the received voltages, the reflectivity factor, Z, and other radar measurements can be inferred. The most commonly used measurement for precipitation estimation is the reflectivity factor. In practice, it has to be measured at a specific polarization state; for NEXRAD, a horizontal polarization state is used.

The measurements of precipitation described above can be elaborated on further for rain. The distributions of raindrop sizes and shapes form the building blocks for deriving physically based rain rate algorithms. Although practical considerations may be just as important, the physical approach provides guidance in developing algorithms for rainfall estimation. The raindrop size distribution (DSD) describes the probability density distribution function of raindrops and can be expressed as

$$N(D) = n_c f_D(D);\ m^{-3} mm^{-1} \tag{3}$$

where $N(D)$ is the number of raindrops per unit volume per unit size interval (D to $D + \Delta D$), n_c is the total drop number concentration, and $f_D(D)$ is the probability density function of raindrop size.

The Z-R algorithm relating the reflectivity factor, Z, to rainfall rate, R, is the most widely used rainfall estimation technique and is the basis of current NEXRAD rainfall estimates. Initially it was used as an engineering solution to the problem of estimation of rainfall from radar. The commonly used Z-R relationships are of the form

$$Z = aR^b \tag{4}$$

In the literature, the coefficients a and b both have been varied to account for such things as variability in the type of rainfall with climatic regions and season. In reality, the relationship of rainfall rate to the concentration and size of raindrops can vary from storm to storm and even between different locations and times within a storm. The radar beam geometry and its relationship to that of the storm can exacerbate these difficulties. Thus, it is not uncommon for raw radar-estimated precipitation rates and accumulations to be in error by 50 percent or more. For example, in cases of flooding from warm rain convective storms in Virginia and Texas, Baeck and Smith (1998) found the radar estimates to be in error by a factor of 3 or more.

continued

BOX 4.1 CONTINUED

However over time, various researchers have tried to provide a physical basis for the *Z-R* relation (see Steiner et al., 2004, for a review). One such approach is the concept of normalized drop size distributions (Sekhon and Srivastava, 1971; Willis, 1984; Bringi and Chandrasekar, 2001; Testud et al., 2001). Using the normalized form of the DSD, *Z* and *R* can be related as

$$\frac{Z}{N_w} = a'\left(\frac{R}{N_w}\right)^b \quad (5)$$

where b is approximately 1.5 and N_W is the normalizing constant defined as the intercept of an equivalent exponential size distribution with the same water content. Equation (5) indicates that most of the fluctuations in *Z-R* relations can be ascribed to variability in N_W. The normalized DSD principle fixes b and accounts for all the variability in rainfall through changing N_W.

Zawadzki (1984) has argued that the measurement procedure can dominate the process of deriving a ground rainfall rate from reflectivity data. Bringi et al. (2003) have studied the variability of N_W over different climatic regimes and conclude that there are significant systematic variations in N_W that could account for changes in the *Z-R* relations. Nevertheless, it is clear that attention to all aspects of the retrieval of rainfall rate from the radar is necessary, and the relative impact of physical processes and practical measurement issues will depend on the prevailing situation.

---

advantage of using radar to estimate precipitation is that measurements can be made over large areas, with fairly high temporal and spatial resolution. In addition, radars can provide rapid updates of the three-dimensional structure of precipitation. Because of these advantages, radar measurements of precipitation have enjoyed widespread use for meteorological applications.

Radar rainfall estimation can be classified broadly into (1) physically based and (2) statistical-engineering-based methods. Physically based rainfall algorithms, as defined here, rely on physical models of the rain medium independent of ground observations, whereas statistical-engineering methods rely on modifications to the algorithm based on the volumetric structure of the radar echoes or the information from related gauge observations. Physically based approaches attempt to solve the inverse electromagnetic problem of obtaining resolution-volume-averaged precipitation estimates from radar backscatter and forward-scatter measurements; they essentially require a polarimetric capability that will become available on NEXRAD in the future (see Chapter 8). Engineering solutions, on the other hand, seek the best possible estimate of rainfall on the ground, recognizing that radar measurements are made aloft by using some feedback mechanism such as gauge data.

Although both physically based techniques and engineering solutions have their role in precipitation measurements, only the engineering approach is presently applicable with NEXRAD. Engineering techniques that focus primarily on accurate estimation of rainfall on the ground range from simple techniques, such as tuning the algorithm coefficients with season or with radar range, to more sophisticated approaches, such as the derivation of nonparametric relationships[3] between the reflectivity factor and the rainfall rate or the use of neural networks. Neural networks in this context refer to computing systems that train themselves from observations to yield ground rainfall estimates from remote sensing observations. A class of hybrid procedures also is evolving that combine the advantages of physically based and statistical-engineering solutions.

The potential applications of quantitative radar precipitation measurements are very broad, including hydrology, agriculture, forestry, studies of the water cycle and water resource management, nowcasting floods, and data assimilation and validation in numerical modeling. Reviewing all of the research findings and issues related to radar rainfall estimation is beyond the scope of this report; we refer the interested reader to discussions by Wilson and Brandes (1979), Austin (1987), Bringi and Chandrasekar (2001), and Krajewski and Smith (2002). Here the focus is on the sources of uncertainty affecting radar rainfall estimates in complex terrain and coastal areas (e.g., Young et al., 1999). One problem with characterizing such remote estimation of rainfall is the lack of systematic strategies and approaches to quantify and characterize the uncertainty structure of the radar rainfall products. Radar rainfall errors can be conceptualized as both systematic (i.e., causing bias) and random, and their statistical determination using ground-based systems requires appropriate approaches that take into account the uncertainty of the reference estimate (Krajewski and Smith, 2002; Habib et al., 2004). In particular, consideration of space-time scale and sample-size effects is important.

[3]Nonparametric relationships refer to algorithms that cannot be expressed in a simple parametric equation, such as $Z = aR^b$, where R is the rain rate, Z is the reflectivity factor (see Box 4.1), and a and b are constants of the algorithm.

The NEXRAD Precipitation Processing System

The NEXRAD Precipitation Processing System (PPS) currently in operation at the NWS consists of five main components: (1) data preprocessing, (2) precipitation rate estimation, (3) accumulation, (4) adjustment, and (5) product generation. Two support functions, precipitation detection and rain gauge data acquisition, are run independently of the PPS and provide additional important information for the main algorithms. The PPS algorithm performance is controlled by 46 adaptable parameters that may be adjusted manually to account for local meteorological conditions such as precipitation type and seasonal climatology, as well as other factors, such as topography, radar siting, or rain gauge network characteristics. However, at present most of these parameters are set identically nationwide. Further details about the individual components are given by Fulton et al. (1998).

The major thrust of data preprocessing is calibration and quality control of the reflectivity data that form the basis for precipitation estimation. Precipitation is estimated on the basis of a so-called hybrid scan that is constructed from the four lowest, fixed-elevation 360° azimuth traverses. Information from the traverse closest to the ground (i.e., the "base scan") is preferred; however, contamination by ground clutter or anomalous propagation echoes may require using data from higher elevations. A decision algorithm is applied to assess data quality within each traverse for selection of the appropriate elevation data. Typically the base scan is used at far ranges—to the extent that this scan clears the terrain (i.e., is not blocked) and does not contain ground or anomalous propagation clutter—whereas a higher-elevation data contribute to the hybrid scan closer to the radar. This hybrid assembly is designed to obtain reflectivity information at an approximately constant altitude close to ground level. The selection of elevation data varies in azimuth depending on local terrain features.

Once the hybrid scan is assembled, radar reflectivity information is converted to rainfall rate using a standard (climatological) *Z-R* relationship. Two options are available—one for a continental and the other for a more tropical environment. Maximum rain rates are capped at approximately 100 mm h^{-1} (4 in. h^{-1}) (150 mm h^{-1} [6 in. h^{-1}] for tropical regions) to avoid unrealistically high values that may result from contamination of the radar data by hail. The radar-based estimates of rainfall rate then are accumulated into hourly totals that are compared to contemporaneous gauge rainfall amounts at the locations of reporting gauges. These pairs of radar and gauge rainfall accumulations are evaluated using a Kalman-filter technique to estimate the bias of the radar rainfall amount, which is subsequently applied to adjust current and future rainfall estimates until a new bias estimate is determined an hour later. Finally, a variety of digital and graphical precipitation products are generated for various operational forecasting uses.

Practical Issues of Estimating Precipitation Using the Reflectivity Factor

As described in Box 4.1, the measured radar reflectivity factor, *Z,* and the rain rate, *R,* are linked physically by the raindrop size distribution, DSD. If radar can measure *Z* sufficiently close to the ground so that the precipitation intensity does not change over this height, the only uncertainty in the transformation from *Z* to *R* arises from the DSD variability. These DSD fluctuations limit the rainfall measurement accuracy to approximately 30–40 percent if a single climatological *Z-R* relationship is used. This accuracy can be improved significantly if the *Z-R* relationship is changed in accordance with the precipitation type. Such estimates of accuracy based on physical considerations are not valid, however, under various practical limitations as described below.

As described earlier, three factors limit the lowest height of the radar measurement: Earth's curvature, beam blockage by the landscape, and ground clutter contamination. Between measurement altitudes and the ground, the rain intensity can change due to further growth, raindrop breakup, or evaporation. If the measurement is above the 0°C isotherm (i.e., the freezing level), the phase of the precipitation changes as well. Contamination by the "bright band," where falling snow, graupel, or hail melts into raindrops and there are changes in the vertical profile of reflectivity, introduces uncertainties that may dominate the errors in radar precipitation estimates. Several methodologies have been used to deal with this problem, such as climatological corrections, range-dependent probability matching and neural network-based rainfall estimates.

The practical problem of operational precipitation estimation involves a number of steps that must be undertaken with care. If radar measurements are well calibrated, the elimination of returns from nonmeteorological targets—foremost, ground clutter—is the first step. Filtering echoes with near-zero velocity during signal processing can mitigate the problem; the effectiveness of this will vary according to the quality of the radar transmitter and the strength of the clutter. In any case, avoidance of the remaining ground or sea clutter echoes at the data processing stage still will be necessary. Algorithms based on Doppler velocity, vertical echo structure, and horizontal echo structure (in that order of importance) can be quite effective in real-time detection of contamination by ground echoes or those resulting from anomalous propagation. Voids created by censoring the contaminated data can be filled by horizontal interpolation (if the void is only a small area) or by vertical extrapolation of noncontaminated data taken at higher elevations. The latter can be done in the same manner as for data from far range where the lowest elevation angles necessarily sample precipitation at an appreciable height above ground.

Next in importance is the removal of echoes contaminated by the bright-band effect. These regions of extremely high reflectivity can be particularly pernicious, even leading to flood warnings in light rain. The height of the 0°C isotherm can be obtained from soundings, model outputs, and more directly, from the vertical structure of radar data themselves. Using all these sources of information is ideal; during frontal passages, the height of the bright band may change rapidly and radar information is crucial. With the 1.0° beamwidth of NEXRAD, the signature of the bright band can be detected clearly up to ranges of 50–70 km, depending on the thickness and intensity of the bright band. Several strategies can be used to overcome bright-band contamination; the simplest is to substitute information from below the bright-band region (if possible) or use the reflectivity of snow above the region and extrapolate the measurement to ground.

Zones of partial beam blocking must be identified for each antenna elevation and a compensation factor applied to the data. This is particularly critical in complex terrain; Chapter 7 provides illustrative examples for the Sulphur Mountain radar. A question can arise as to whether it is better to use non-blocked data from higher-elevation beams or to compensate for the blocked fraction of the lowest beam.

Once the decontamination and correction of data are completed, extrapolation to the ground must be done. The broadening of the radar beam with range, the minimum beam elevation angle, and Earth's curvature all serve to increase the distance between the radar scattering volume aloft and the surface. As this distance increases with range, it often is referred to as the range effect. Here the removal of the bias associated with the vertical profile of reflectivity (VPR) is necessary. This profile can be estimated from observations at short ranges, where the scans reach down to the surface or near the surface. However, the VPR can be highly variable in time and space, so that the VPR determined at short ranges may be quite different from that at far ranges; a conservative approach is preferable. At short ranges, where voids were created by removal of ground-contaminated data, the nearby VPR should be sufficient for an effective extrapolation.

The Z-R relationships also can be obtained by a direct radar-gauge comparison. This can be done in various ways. One commonly used procedure is via regression of synchronous radar and gauge measurements. Alternative approaches have been suggested for developing observation-based mapping from radar observations to rainfall on the ground. One such procedure is the neural network technique, which provides a mechanism to build a nonparametric relation between surface rainfall and the vertical profile of reflectivity aloft (Liu et al., 2001).

THE EVOLVING NEXRAD SYSTEM

The NEXRAD system configuration is not static, but rather continues to evolve through an ongoing NEXRAD Product Improvement (NPI) Program. Stated objectives of this program (Saffle et al., 2001) are to

- ensure the capability to implement advances in science and technology to improve forecasts, watches, and warnings;
- minimize system maintenance costs; and
- support relatively easy upgrades in technology so that a large-scale NEXRAD replacement program may be indefinitely postponed.

The NPI Program provides a means for introducing continuing improvements in science and technology into the NEXRAD system on an ongoing basis. The NPI Program works to develop and introduce system improvements in an orderly and seamless manner. Thus, the NEXRAD system a decade hence will be substantially improved over that of today.

The current NPI Program emphasizes two major thrusts. One is to replace the data acquisition and processing systems in the original NEXRAD with open-system hardware and software. This development increases the overall capability of the NEXRAD system for data acquisition, processing, display, dissemination, and archiving; facilitates implementation of new algorithms for processing radar data; and reduces costs for system operation and maintenance. Field deployment of the ORPG component, which executes the NEXRAD algorithms and produces the image products, was completed in FY 2002. The ORPG improvements provide a capability for (1) data quality improvements, such as AP clutter detection and suppression and identification of nonprecipitation echoes; (2) new polarimetric-based products such as improved precipitation estimation and hydrometeor particle identification; and (3) new products that may be assimilated directly into operational numerical models. Introduction of the second component, the open RDA (ORDA) unit, is in progress. The ORDA improvements will include (1) availability of a modern Doppler spectral processing platform, including digital receivers, for improved data fidelity; (2) a provision for range-velocity ambiguity mitigation techniques using phase coding and dual-pulse repetition interval waveforms; (3) the capability for polarimetric sensing and processing to more accurately measure hydrometeor properties, such as drop size distributions and precipitation phase; and (4) custom VCPs to allow site-specific volume scans adapted to local weather needs. These improvements are expected to be operational in the next few years. The third major component of the original NEXRAD, the PUP display unit, has been replaced with open-architecture systems in different ways by the three major NEXRAD user agencies.

The second major thrust of the NPI Program is directed toward introduction of a polarimetric capability for NEXRAD. Such a capability could provide improved precipitation measurements as well as new capabilities for identifying hydrometeor types (e.g., recognizing the presence of hail, discriminating between rain and snow regions) and enhanced ability to screen out artifact echoes such as those caused by ground clutter or birds. The polarimetric capability is scheduled for implementation in the next few years.

The NPI Program is not limited to modifications of NEXRAD itself. An enhanced software environment, termed Common Operations and Development Environment (CODE), is being provided to facilitate use of the open-architecture system capabilities and linkage of the NEXRAD data to other agency weather data systems such as AWIPS. Data from other radar systems also can augment the NEXRAD dataset, and plans and procedures are being developed to incorporate data from appropriate FAA and other radar systems. These include the FAA's Terminal Doppler Weather Radar (TDWR) and short- and long-range surveillance radars (Air Route Surveillance Radar [ARSR] systems and Airport Surveillance Radar [ASR]) (see Chapter 8), as well as atmospheric wind profilers (Rich, 1992) and radar systems operated by television stations and other private entities. This development will enhance the available coverage and also the backup capabilities in case of a NEXRAD outage.

The national coverage, improved accuracy, and rainfall estimation capabilities of the NEXRAD system have advanced the practice of hydrologic forecasting and water resource management. If dual-polarization capabilities are incorporated into the current system as planned, further improvements in precipitation measurements will occur, particularly in the quantification of high-intensity rainfall rates and the characterization of snowfall. However, closer radar spacing would be needed for representative near-surface coverage throughout the continental United States. The latter implies the need for an affordable means of dealing with the inherent inability of any widely spaced network to provide comprehensive near-surface surveillance over large portions of the country.

SITING OF THE NEXRADS

In preparation for installation of the national network of NEXRADs, surveys were performed nationwide by SRI International, Inc., and Metcalf & Eddy, Inc., to evaluate potential radar sites. A three-step process was employed by the survey team to site the NEXRADs. First, an Initial Site Assessment was done, which entailed an examination of a potential site by the analyses of maps and other data without visiting the site. Then a Preliminary Site Survey was conducted by visiting the site to assess the potential radar coverage and the ability of the site to fulfill the requirements of the three principal users (i.e., FAA, NWS, DoD). A report was prepared for the DOC as a result of the preliminary siting. Finally, based on a favorable preliminary site survey, an In-Depth Site Survey was performed, which included a second site visit and another report for the DOC containing updated information and a detailed analysis of the site and radar coverage. Estimated site preparation costs, anticipated environmental and historical impacts caused by site construction and future radar operations at the site, and measurements of electromagnetic frequencies in the area were included in the report.

General Siting Considerations

To select the best sites for the NEXRADs, the DOC outlined four key factors that were important in the selection of sites, including (1) optimization of radar location for an unobstructed view in the prevailing direction of approaching hazardous weather; (2) minimization of ground clutter over the areas of greatest interest; (3) minimal impact on the environment and electromagnetic interference; and (4) no degradation of existing radar coverage over coastal regions. More detailed information on siting weather radars is provided by Leone et al. (1989), Doviak and Zrnic (1985), and the *NEXRAD Siting Handbook* (NEXRAD Joint System Program Office, 1983).

Site selection was based on providing maximum radar coverage in the priority areas required by the NWS, FAA, and DoD over highly populated regions while keeping installation and operational costs at a minimum. Both the NWS and the FAA require radar coverage over large areas, the former needing coverage over high-population areas to detect severe weather, particularly at low altitudes, and the latter needing coverage of both airport terminal areas and airways en route. The NWS priority regions of coverage were based on climatological records of the frequency of occurrence and prevalent locations of severe weather in the area of interest. Outside the priority coverage areas, NEXRAD Doppler radars can detect most types of hazardous weather out to a range of approximately 148 km. The DoD requires radar coverage within 65 km of its highest-priority military and civilian facilities. Weather radar sites in existence prior to the installation of NEXRADs were given first consideration as potential sites, but they did not always meet the requirements outlined by the DOC; this was the case for the Sulphur Mountain radar (see below).

An important consideration in the siting of a NEXRAD was its contribution to the national network of NEXRAD coverage. Overlapping coverage of NEXRADs can be crucial for the early detection of severe weather from directions not well observed by the closest NEXRAD. An additional criterion for the network of NEXRADs was the ability to provide coverage of FAA airways from 1.83 km (6000 ft) AGL to 21.34 km (70,000 ft) MSL over the continental United States east of the Rocky Mountains and over the San Francisco-San Diego Corridor (to the Sierra Nevada Mountains) and 3.05 km (10,000 ft) AGL to 21.34 km (70,000 ft) MSL elsewhere.

NEXRAD Site Surveys for Los Angeles, California (Sulphur Mountain Site)

Initial, preliminary, and in-depth site surveys were performed for the Los Angeles, California, area during 1986–1987. A NWS Weather Surveillance Radar-1974 C-band (WSR-74C) radar was located in the Los Angeles basin at the time. Due to the severe ground clutter return from the surrounding metropolitan area, the WSR-74C site was not considered a favorable location for the installation of NEXRAD. Furthermore, AP effects, resulting from interaction of the radar beam with the coastal marine inversion layer that frequently persists in Los Angeles and surrounding metropolitan areas (see Chapter 5), were known to be quite severe. Given these limitations in radar performance and the NWS requirement of priority radar coverage down to 0.61 km (2000 ft) AGL to detect severe weather in the prevailing approach zones (Figure 4.2), an elevated site was considered critical to minimize the effects of ground clutter and AP.

Several candidate sites were examined including Sulphur Mountain, Saddle Peak, and Castro Peak. In response to a request by the NWS Office of Hydrology, these sites were visited more than once to address their suitability in providing the best radar coverage possible in a region of com-

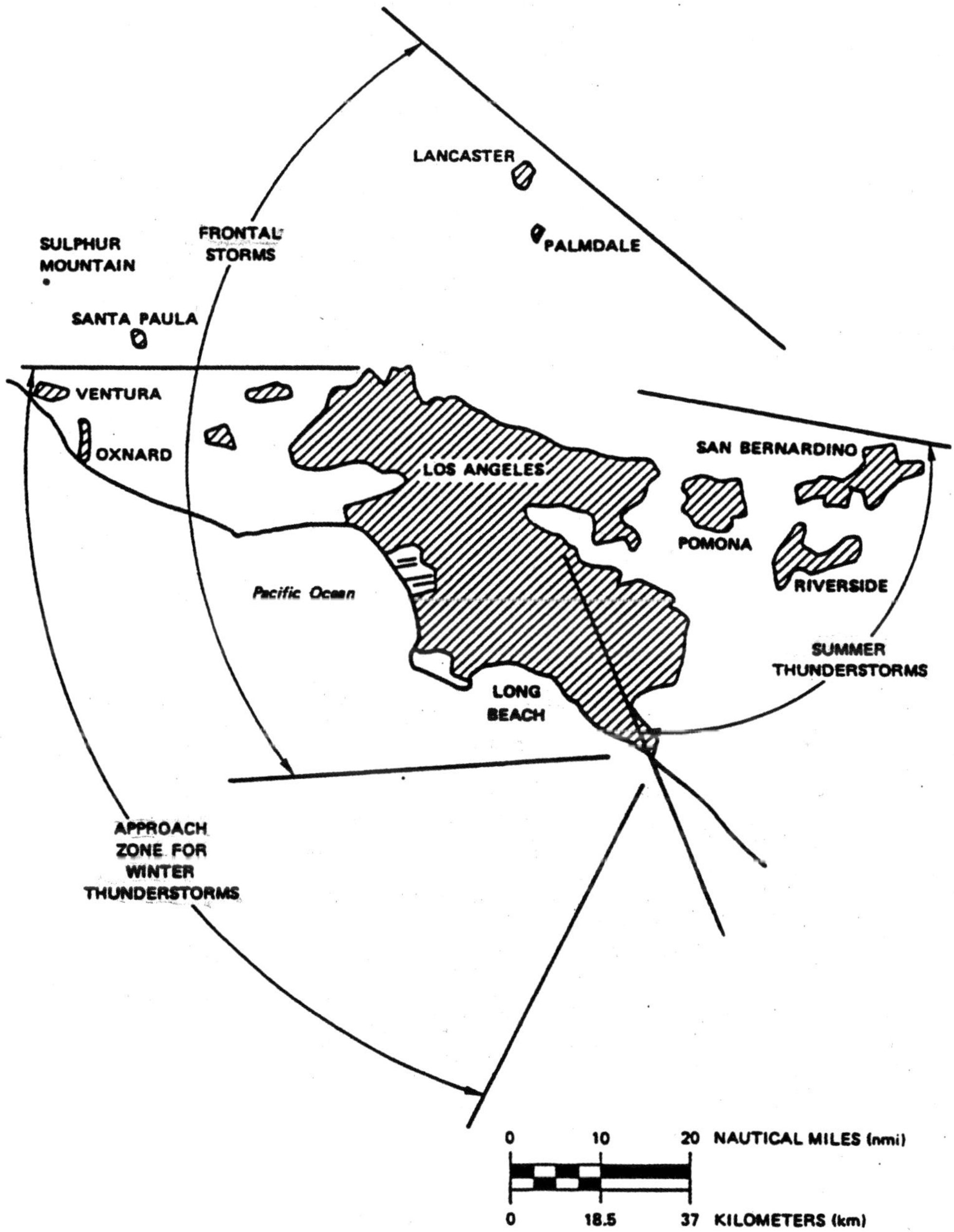

FIGURE 4.2 The Sulphur Mountain radar site shown with respect to the prevailing area from which precipitating systems approach and to the cities of Los Angeles, Ventura, and Oxnard. SOURCE: SRI International, Inc.

plex terrain. Saddle Peak was eliminated from consideration due to the existence of several communication towers that were already in place, the performance of which would have been negatively affected by the installation of NEXRAD. Castro Peak was abandoned as a potential site because its elevation was not high enough to keep the main radar beam above the climatological range of variability of inversion heights, and thus the integrity and utility of the radar data would be diminished by the frequency of AP effects. The Sulphur Mountain site did not present these problems, and eventually it was selected by the survey team as the best site to meet the priority radar coverage requirements. This site is located 78 km northwest of downtown Los Angeles, 22 km north of the NWS Los Angeles-Oxnard WFO, and 19 km northeast of downtown Ventura and the town of Ojai (see Chapter 6). The site is at the top of Sulphur Mountain at an elevation of 0.83 km (2726 ft) MSL (Figure 4.3). A series of ridges blocks radar coverage to the north, but the site survey indicated that the radar would be well positioned to provide good coverage over Los Angeles and surrounding communities (Figure 4.4) out to the 148-km range and over potential flooding regions at the base of the San Gabriel Mountains (Leone and Johnson, 1986).

Using selected clutter models, the survey team examined degradation of radar coverage due to ground clutter for the Sulphur Mountain site. It was anticipated that the radar would be able to reliably detect weather beyond the 3.7-km to 9.3-km range. The overall intensity of ground clutter return was expected to be much lower than that observed at the WSR-74C radar site. Adjustments in the pulse-repetition frequency (PRF) used by the radar were expected to improve weather coverage in a particular area of interest by minimizing the overlap of long-range—that is, greater than 148-km—radar echoes with ground clutter areas.

Because the minimum elevation angle of the main beam is 0.5°, the current institutional limitation, the center of the radar antenna would have to be located at a height at least level with the average height of the nearby hills in the directions of interest. The necessary height of the radar tower was determined to be approximately 24 m (81 ft), thus a standard 20-m tower was proposed (Burns et al., 1987). At an elevation of 0.5°, the height of the center of the radar beam at a range of 148-km will be approximately 2.57 km (8432 ft) above the site level. Figure 4.5 provides an illustration of this concept graphically, but at a range of approximately 120 km from the radar, where the height of the center of the beam at 0.5° elevation above site level is 1.91 km (6266 ft) (i.e., the difference between the 2.74-km [8990-ft] height of the beam above MSL at 120 km range and the 0.83-km [2726-ft] height of the Sulphur Mountain site). Thus, the elevation of the

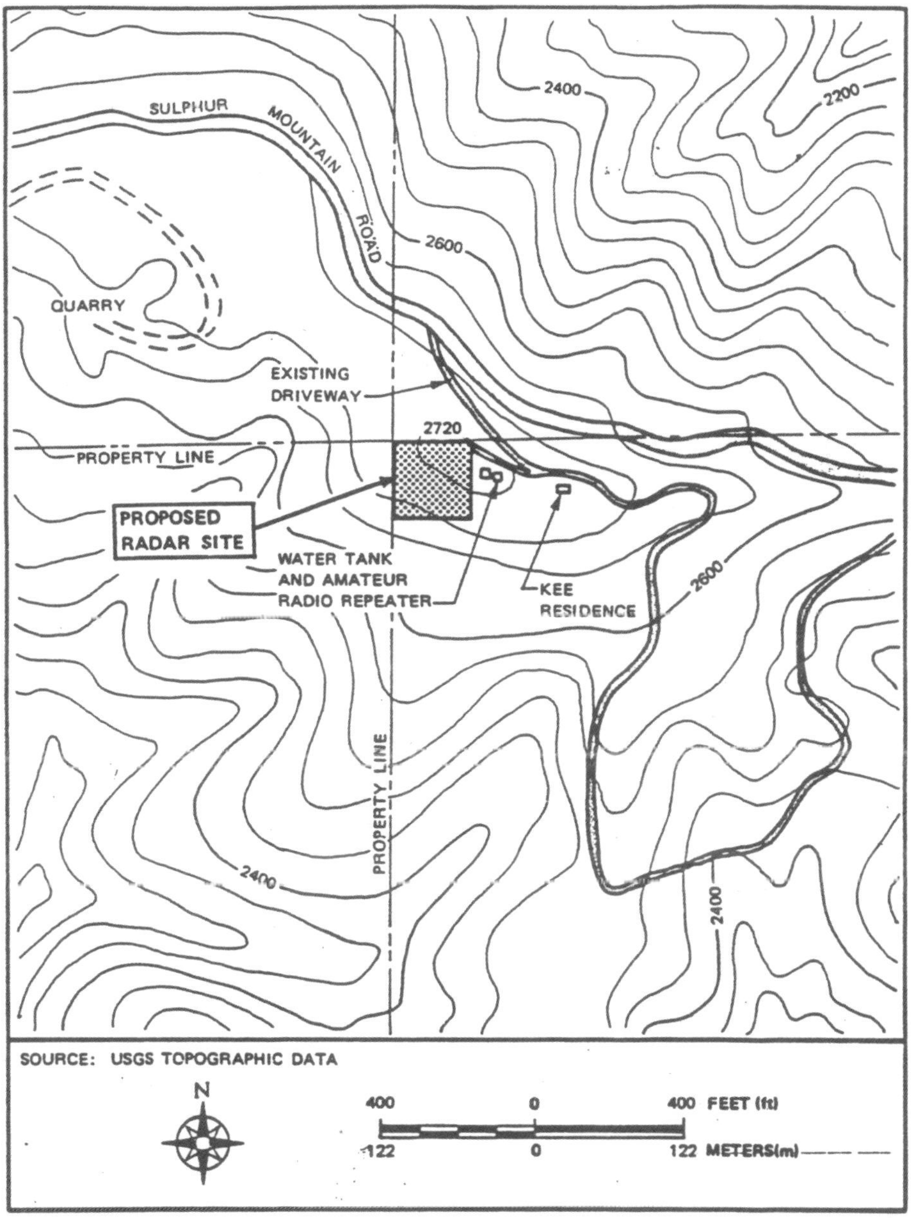

FIGURE 4.3 Topographical map showing the location of the Sulphur Mountain radar site at an elevation of 0.83 km (2726 ft). The site elevation was initially estimated at 2720 ft, as shown in this figure, but it was later updated to 2726 ft. SOURCE: SRI International, Inc.

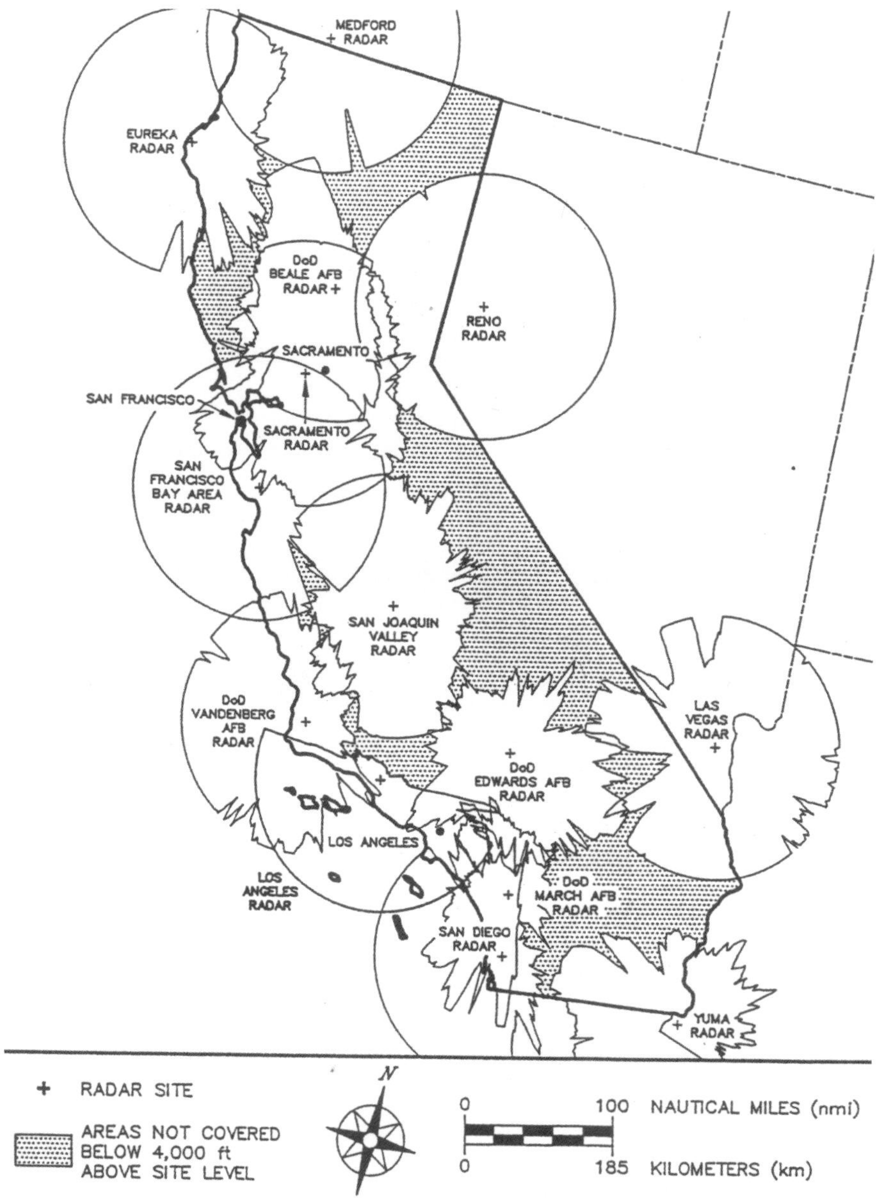

FIGURE 4.4 Radar coverage 1.22 km (4000 ft) above site level for the Sulphur Mountain radar (i.e., covering the Los Angeles area) and for the immediate surrounding radars, including Vandenberg Air Force Base (AFB), Edwards AFB, and San Diego. Note that the March AFB radar has since been replaced with the Santa Ana radar. SOURCE: SRI International, Inc.

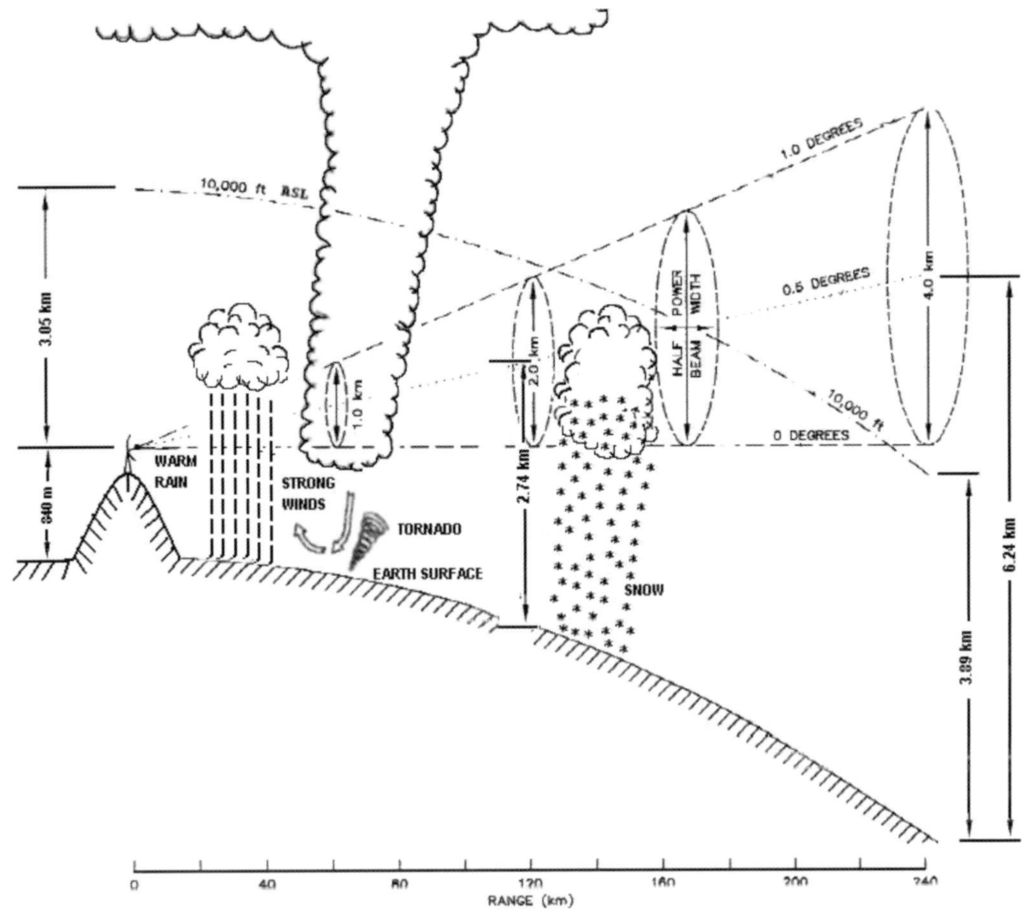

FIGURE 4.5 Cartoon depicting the Sulphur Mountain radar, at an elevation of 0.83 km (2726 ft) and with a 20-m tower, transmitting at the current minimum beam elevation of 0.5°. The cartoon shows how the distance between the beam scan and Earth increases with distance from the radar due to Earth's curvature and the rising of the beam. The 3.05 km (10,000-ft) ASL (above site level) swath also is shown. SOURCE: Modified by V. Chandrasekar from SRI International, Inc.

proposed Sulphur Mountain site was deemed sufficient to provide radar coverage above the strong inversion layer that often covers the Los Angeles area.

After completion of the preliminary and in-depth site surveys, construction at the site began on November 15, 1993. Work included clearing of the site and placement of forms for the concrete foundation. The construction crew returned to the site on November 28, 1993, to remove the forms, and the Sulphur Mountain radar was erected on December 16, 1993. A complete chronology of events leading up to the siting of the Sulphur Mountain NEXRAD can be found in Appendix C.

5

Observational Challenges of Low-Level Radar Coverage in Complex Terrain and Coastal Areas

Regions of complex terrain present a challenging environment for using NEXRAD to observe potential flash flood-causing weather systems. Two of the more serious problems are those of beam blockage and beam overshoot. The scanning radar beam can be partially or completely blocked for various directions when the radar is sited in an area surrounded on some sides by higher terrain (Westrick et al., 1999; Young et al., 1999). Placing the radar at a higher elevation or on a mountaintop may alleviate this problem, but doing so may cause the radar beam to overshoot lower levels. As a result, the precipitation intensity can be underestimated, or some precipitation may go undetected altogether (Brown et al., 2002; Wood et al., 2003a).

NEXRADs sited near the coast are especially valuable for providing observations over the ocean where little other data are available. This is particularly true for the west coast of the United States, where weather systems approach predominantly from the west. However, the observational challenges posed by complex terrain are compounded for coastal regions because of the presence of a large water body that affects the atmospheric conditions. Variations in atmospheric temperature, moisture, and pressure can significantly alter the propagation of a radar signal, affecting the radar's ability to detect and quantify precipitation. Deploying a radar along a coastal mountain range must be a compromise between providing offshore low-level coverage and avoiding negative effects of atmospheric conditions on radar signal propagation. Low-level coverage is important for weather systems approaching the coastal mountains of the West during the winter and early spring, because these systems tend to be shallow yet still produce significant rainfall intensities and amounts. By the time rainfall is detected and measured by in situ rain and stream gauges, there can be little time to disseminate river and streamflow forecasts and flash flood warnings. Therefore, advanced detection of approaching weather systems by radar plays a critical role in flash flood prediction. These observational challenges posed to NEXRAD in complex terrain and coastal regions—for example, in Southern California's Ventura and Los Angeles Counties—are discussed in more detail in this chapter.

BEAM VISIBILITY

Propagation of radar waves in the atmosphere is a well-understood process (Bean and Dutton, 1966). If the atmospheric properties such as temperature, humidity, and pressure are known, the propagation path of the radar beam is easy to characterize. Design studies of radar networks usually assume a standard atmosphere (Doviak and Zrnic, 1985). Assuming that the beam propagates under standard atmospheric conditions allows determination of the potential interactions of the beam with the surrounding terrain. To achieve this, two other factors must be considered. First, the radar beam spreads as it propagates away from the radar so that the NEXRAD beam is more than 1 km wide at a distance of about 60 km from the radar site. Second, the power distribution within the radar beam is not uniform; most of the power is concentrated near the main axis of the beam, and the intensity falls off approximately following a Gaussian pattern. Thus, to properly determine the extent to which terrain affects radar observations, detailed information about the terrain, location, and hardware parameters of the radar and its scanning pattern (i.e., antenna elevation angle) must be known in addition to atmospheric conditions.

Complete blockage of the radar beam by terrain features occurring at a given distance and direction prevents detection of weather echoes past the location of the blockage. The blocking element (i.e., hills, mountains, nearby buildings) leads to a strong and permanent echo that is visible on the radar display. If the radar beam is only partially blocked, it may still be useful for observing weather echoes if the amount of power remaining in the beam is large enough for the radar return to be above the signal detection level. However, precipitation observations collected by partially blocked radar beams may lead to biased results if left uncorrected. Corrections for partial beam blockage can be part of radar data processing algorithms, but correcting the data often is difficult because the degree to which the beam is obstructed rarely is known accurately (Seo et al., 2000; Kucera et al., 2004). In addition, the corrections may rely on assumptions that are difficult to verify because complex terrain modifies observed echo patterns (Warner et al., 2000; Ralph et al., 2003).

There are two approaches to evaluating beam blockage patterns for a given location. The first approach is based on using terrain elevation data, such as those available from digital elevation models (DEMs). For the United States, digital elevation data are available with a spatial resolution of about 30 m and in some locations even 10 m. (Airborne lidar technologies exist to provide such data with a resolution of 1 m.) Knowing the topography, radar location, and parameters such as beamwidth and antenna pointing direction, and assuming standard atmospheric conditions, one can calculate the power loss in the radar beam along its propagation path (Kucera et al., 2004). Chapter 7 provides more information about this method (including a schematic representation) and shows specific calculations for the Sulphur Mountain radar and adjoining radars.

The second approach is based on actual data collected by radar. One can calculate the spatial patterns of the probability of detection (POD) of an echo as the ratio of the number of returns exceeding a certain threshold of radar reflectivity (e.g., 20 dBZ) and the total number of scans. If such calculations are based on a sufficiently large dataset, a clear pattern will emerge. Backscatter from the terrain that completely blocks the beam will be detected all the time, and beyond these obstacles there will be echo voids. For unblocked areas, POD values will correspond to the climatological frequency of rainfall (about 5 percent for most of the United States); partial beam blockage will result in values somewhere in between. This last result may come as a surprise, but it is a consequence of the fact that the propagation conditions usually deviate from those of the standard atmosphere. The advantage of the DEM-based procedure is that it can be used in the network planning stage, but its weakness is the assumption of particular (e.g., standard) propagation conditions. On the other hand, the radar data-based analysis shows the actual patterns of radar coverage, but it requires processing large amounts of data and may be attempted only after the radar has been deployed for a significant amount of time to collect data.

PROPAGATION OF ELECTROMAGNETIC RADAR SIGNALS

As noted above, the propagation of microwave signals, such as those transmitted by radar, is influenced by atmospheric conditions. Deviations of the spatial distribution of moisture, temperature, and/or pressure from the standard atmosphere may lead to nonstandard or anomalous propagation (AP) of radar signals, perturbing the position in space illuminated by the radar as well as the quality of data obtained. Anomalous propagation conditions can lead to significant contamination of radar data from ground echoes normally not seen by the radar. Such echoes must be recognized and acknowledged for operational applications such as precipitation nowcasting and flash flood warning.

Radar reflectivity due to anomalous propagation may resemble that of precipitation (e.g., Steiner and Smith, 2002, and references therein). Contamination of radar reflectivity from the ground due to anomalous propagation is limited primarily to the lowest scan or two (e.g., below 2° elevation). These echoes may be recognized easily by a well-trained person on the basis of the great spatial variability of the reflectivity and near-zero radial Doppler velocity of the AP echoes. This is particularly true for anomalous propagation echoes over land, but less so over water. Nonetheless, the development of automated procedures for radar data quality control to deal with AP echoes remains a challenge today (Alberoni et al., 2001; Krajewski and Vignal, 2001; da Silveira and Holt, 2001; Charalampidis et al., 2002; Steiner and Smith, 2002). Many locations across the United States experience atmospheric conditions leading to anomalous signal propagation on a regular basis, as highlighted by Steiner and Smith (2002). Coastal areas, including southern California, are particularly prone to this problem. For other areas, anomalous propagation conditions may be limited to nighttime or vary by season.

Refractive Index and Signal Propagation

A commonly used quantity to describe the propagation behavior of electromagnetic signals is the index of refraction or refractivity (Box 5.1). The vertical gradient of the refractivity within the lowest several hundred meters above the ground is especially important for characterizing radar signal propagation (e.g., Pratte et al., 1995). The normal decrease in atmospheric refractivity with altitude tends to bend the radar rays downward so as to extend coverage beyond that expected with a uniform atmosphere. There are four basic modes of propagation, including subrefraction, normal refraction, superrefraction, and trapping or ducting (Box 5.1). The abnormal propagation of electromagnetic waves is called anomalous propagation.

Trapping or ducting is the most severe case of anomalous signal propagation and results in ground returns (AP echoes) from distant locations where the radar beam intersects the ground or objects at Earth's surface (Figure 5.1). In order to propagate energy within the duct, the angle that the radar beam makes with the duct should be small, usually less than a degree; thus, only the lowest-elevation scans of surface-based radar are affected.

BOX 5.1

Index of Refraction and Ducting

The index of refraction, n, or the refractivity, N, can be approximated by

$$(n-1)\cdot 10^6 = N = \frac{77.6p}{T} + \frac{3.73\cdot 10^5 e}{T^2} \quad (1)$$

where p is the barometric pressure in millibars, e is the partial pressure of water vapor in millibars, and T is the absolute temperature in degrees Kelvin (Bean and Dutton, 1966; Gossard, 1977; Skolnik, 1980; Babin, 1996; Fabry et al., 1997). The decrease of refractivity with height can be defined as $\frac{dN}{dh}$, and the value of this quantity is used to determine the mode of propagation, where

- subrefraction occurs when $\frac{dN}{dh} > 0\ \mathrm{m}^{-1}$;
- normal refraction occurs when $0 > \frac{dN}{dh} > -0.0787\ \mathrm{m}^{-1}$;
- superrefraction occurs when $-0.0787 > \frac{dN}{dh} > -0.157\ \mathrm{m}^{-1}$; and
- trapping or ducting occurs when $\frac{dN}{dh} < -0.157\ \mathrm{m}^{-1}$.

A simplified approximate model of propagation in atmospheric ducts (Skolnik, 1980) predicts a maximum wavelength, λ_{max}, that can be propagated in a surface duct of depth, d, as given by

$$\lambda_{max} = 2.5d^{\frac{3}{2}}\sqrt{-\frac{dn}{dh}} \quad (2)$$

where λ_{max}, dh, and d are in the same units (e.g., meters).

Only those radar beams launched nearly parallel to the duct will be trapped. However, often only parts of the radar beam may be trapped (Figure 5.2). Atmospheric ducts generally are tens to hundreds of meters thick (Gossard, 1977; Cook, 1991; Babin, 1996; Brooks et al., 1999). Box 5.1 presents a simplified model for estimation of the maximum wavelength that can be propagated in a duct depending on its depth. For an operational NEXRAD with a wavelength of 10 cm (S-band), the duct must be at least 22 m thick for trapping to occur (Heiss et al., 1990; Baer, 1991; Crum et al. 1998).

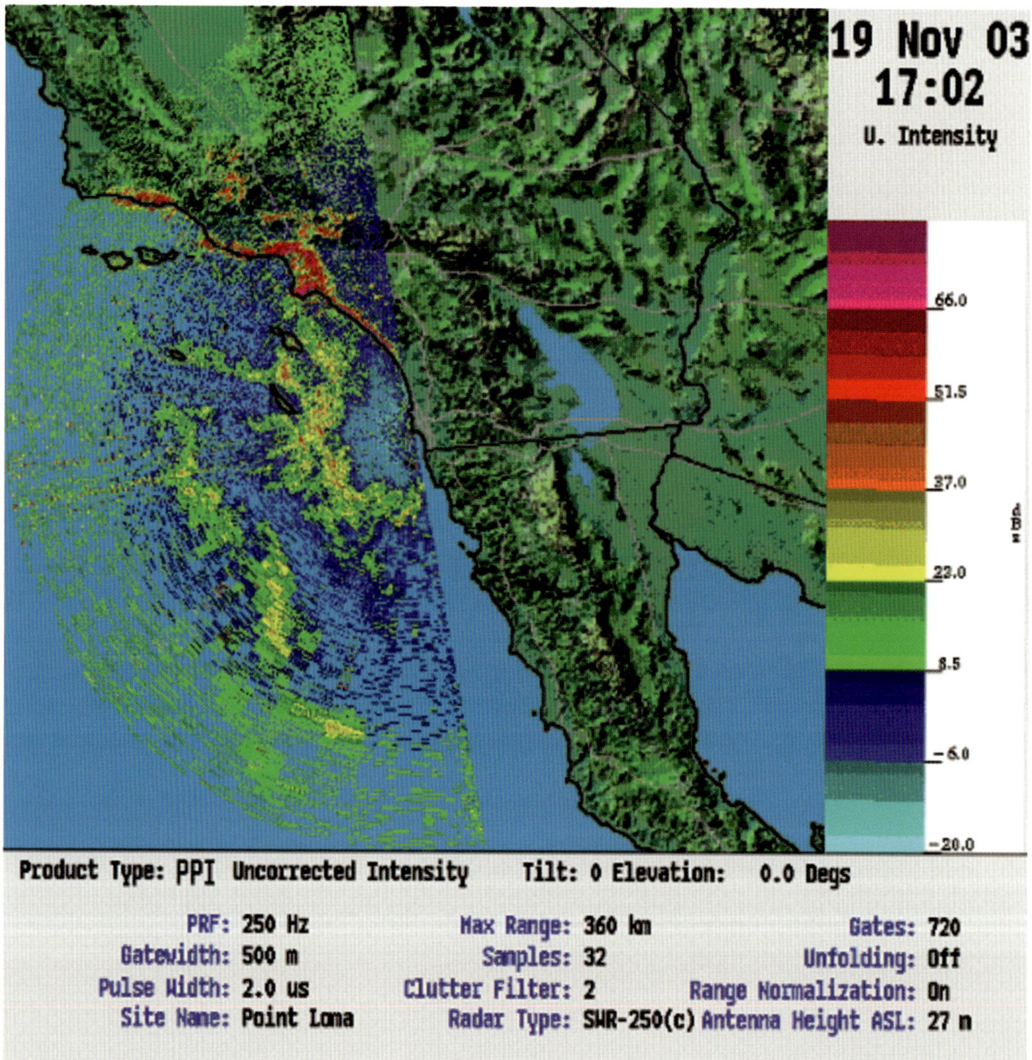

FIGURE 5.1 Sea and ground clutter radar echoes due to atmospheric ducting effects as seen by a C-band Doppler Supplemental Weather Radar (SWR) positioned in San Diego on November 19, 2003. SOURCE: Richard Paulus, Space and Naval Warfare Systems.

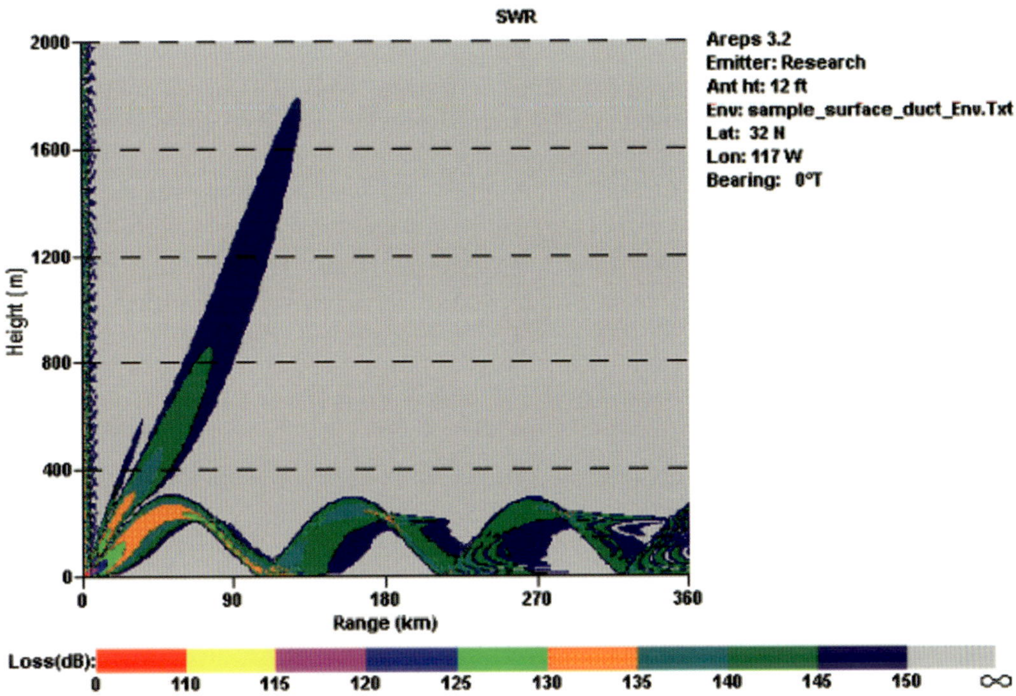

FIGURE 5.2 Numerical simulation of radar signal propagation under surface ducting conditions with parts of the radar beam's signal being trapped. SOURCE: Richard Paulus, Space and Naval Warfare Systems.

A duct is produced when the index of refraction decreases rapidly with height. To achieve this, the temperature must increase and/or the humidity (water vapor content) must decrease with height. Temperature inversions must be very pronounced to produce superrefraction, while water vapor gradients are more effective than temperature gradients alone (Fabry et al., 1997). A common cause of ducting is the movement of warm dry air from land over cooler bodies of water (e.g., ocean) causing a temperature inversion in the boundary layer. At the same time, moisture is added by evaporation from the water surface, producing a moisture gradient (Skolnik, 1980; Puzzo et al., 1989). Evaporation ducts are common just above the surface of the sea, where air may become saturated by evaporation from the sea surface. Over land, ducting often is caused by radiational cooling during clear nights, particularly in summer when the ground is moist. Thus, over land, ducting is most noticeable at night and tends to disappear during the warmest part of the day (e.g., Moszkowicz et al., 1994). Superrefraction or ground ducts also may be produced by the diverging downdraft under a thunderstorm and resulting gust fronts. The relatively cool air, which spreads out from the base of a thunderstorm, may produce a temperature inversion within the lowest layer, possibly several hundred meters thick. The moisture gradient along the outflow boundary also is appropriate for the formation of a duct. The conditions favorable for the formation of a thunderstorm duct are relatively short-lived and have time scales on the order of 30 minutes to an hour, although in extreme cases such conditions may last for hours (Weber et al., 1993).

Climatological Assessment of Signal Propagation for the Southwestern United States

Using a 16-year record of operational sounding data (1973–1988), the potential for anomalous propagation conditions to occur throughout the continental United States was assessed climatologically by Steiner and Smith (2002). Similar studies, for example, have been conducted by Bech et al. (2000, 2003) using soundings for Mediterranean coastal sites, Babin (1996) using helicopter-based refractivity measurements off the coast of Wallops Island (Virginia), and Gossard (1977) using air mass analyses. For each sounding, Steiner and Smith (2002) determined the average refractivity gradient in the lowest 500 m (1640 ft) above ground level (AGL). (The maximum gradient might be more relevant for the radar signal propagation problem; however, the limited and variable vertical resolution of the operational sounding data may result in questionable maximum gradient values.) These values were then compiled into a climatology of average refractivity gradients for each operational sounding station and used to study the likelihood of atmospheric conditions across the United States that are susceptible to anomalous propagation of radar signals. The sounding-based climatology highlights large-scale temperature inversions and moisture gradients, yet only by chance does it capture conditions favorable to anomalous propagation produced by thunderstorm outflow boundaries. Moreover, relatively thin layers of strong vertical gradients may cause anomalous propagation, but the operational sounding data (variable resolution of one to several hundred meters) do not resolve tens of meters in the vertical. Alternatively, high-resolution refractivity profiles may be obtained, for example, from detailed atmospheric measurements aboard an airplane (e.g., Babin, 1996, used a helicopter) or combined profiler and radio acoustic sounding systems (RASS) (Gossard et al., 1995). Lidar-based observations of temperature and water vapor (e.g., Eichinger et al., 1993, 1999) may also provide a basis for obtaining high-resolution refractivity information. In addition, Fabry et al. (1997) and Fabry (2004) showed that radar-based phase measurements of ground targets can be used to reveal the spatial, near-surface structure of atmospheric conditions and thus the index of refraction.

Many areas of the continental United States have conditions that can create anomalous propagation on a regular basis (Steiner and Smith, 2002). The southwestern part of the United States, particularly Southern California, is especially prone to this problem (Gossard, 1977; Pappert and Goodhart, 1977; Babin and Rowland, 1992; Burk and Thompson, 1997). Significant local, regional, diurnal, and seasonal differences are found among the different sounding stations, as highlighted in Figures 5.3 and 5.4. Figure 5.3 shows analyses based on the 12 UTC (coordinated universal time) soundings, reflecting nighttime conditions ranging from 4 a.m. LST (local standard time) on the West Coast to 7 a.m. (about sunrise) at the East Coast. Figure 5.4 is based on the 00 UTC soundings, highlighting conditions in the late afternoon, ranging from 4 p.m. LST on the West Coast to 7 p.m. (about sunset) on the East Coast. Both figures show the distribution of average refractivity gradients within 500 m (1640 ft) AGL for each operational sounding station. The critical categories of "superrefraction" and "trapping" are shaded in gray and black, respectively, and the fraction of the box covered indicates the percentage of soundings exhibiting such atmospheric conditions.

The general nighttime pattern revealed by Figure 5.3 is that conditions of superrefraction may occur anywhere in the southwestern part of the United States. During the winter season (Figure 5.3a), superrefractive propagation conditions occur most likely south of 40°N latitude, particularly along the coastlines. For those areas, on average, superrefractive conditions may occur at least once a week and may be two or three times for Southern California. The maximum likelihood of superrefractive conditions is displayed during the summer (Figure 5.3c), when chances for such conditions to occur exceed 20 percent throughout most of the United States and are greater than 30 percent for Southern California and most of the eastern seaboard states (see Figures 1 and 2 of Steiner and Smith, 2002). The spring (Figure 5.3b) and fall seasons (Figure 5.3d) also show widespread conditions of superrefraction at least once a week, with local maxima in excess of 30 percent generally south of 40°N latitude and throughout states bordering the ocean. Significant trapping conditions (at least once a week) occur

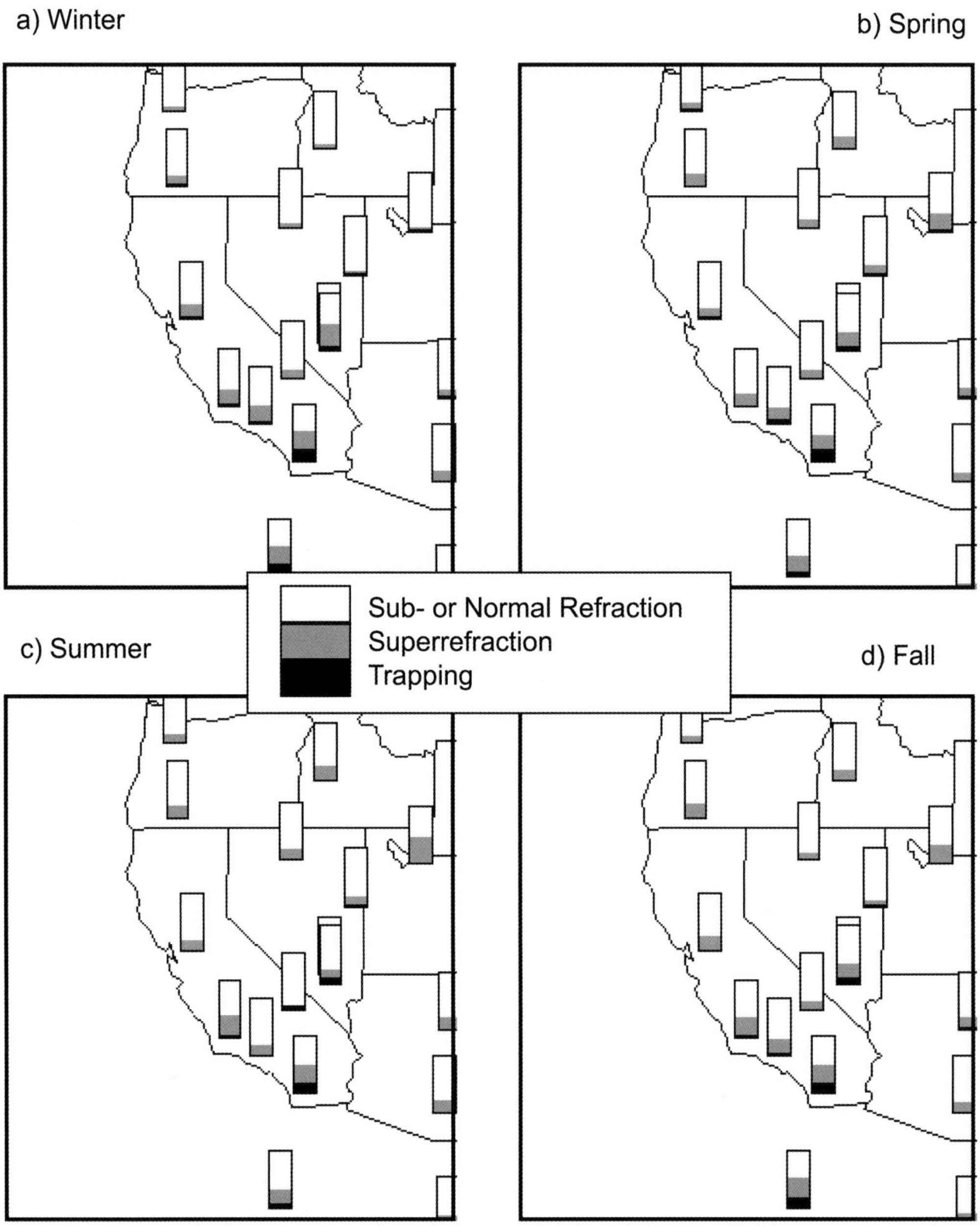

FIGURE 5.3 Average refractivity gradient within 500 m AGL based on a 16-year climatology of operational soundings taken at 12 UTC and shown by season. The fraction of a box shaded either gray (superrefraction) or black (trapping) indicates the percentage of soundings of that particular station exhibiting the respective atmospheric conditions. SOURCE: Adapted from Steiner and Smith (2002).

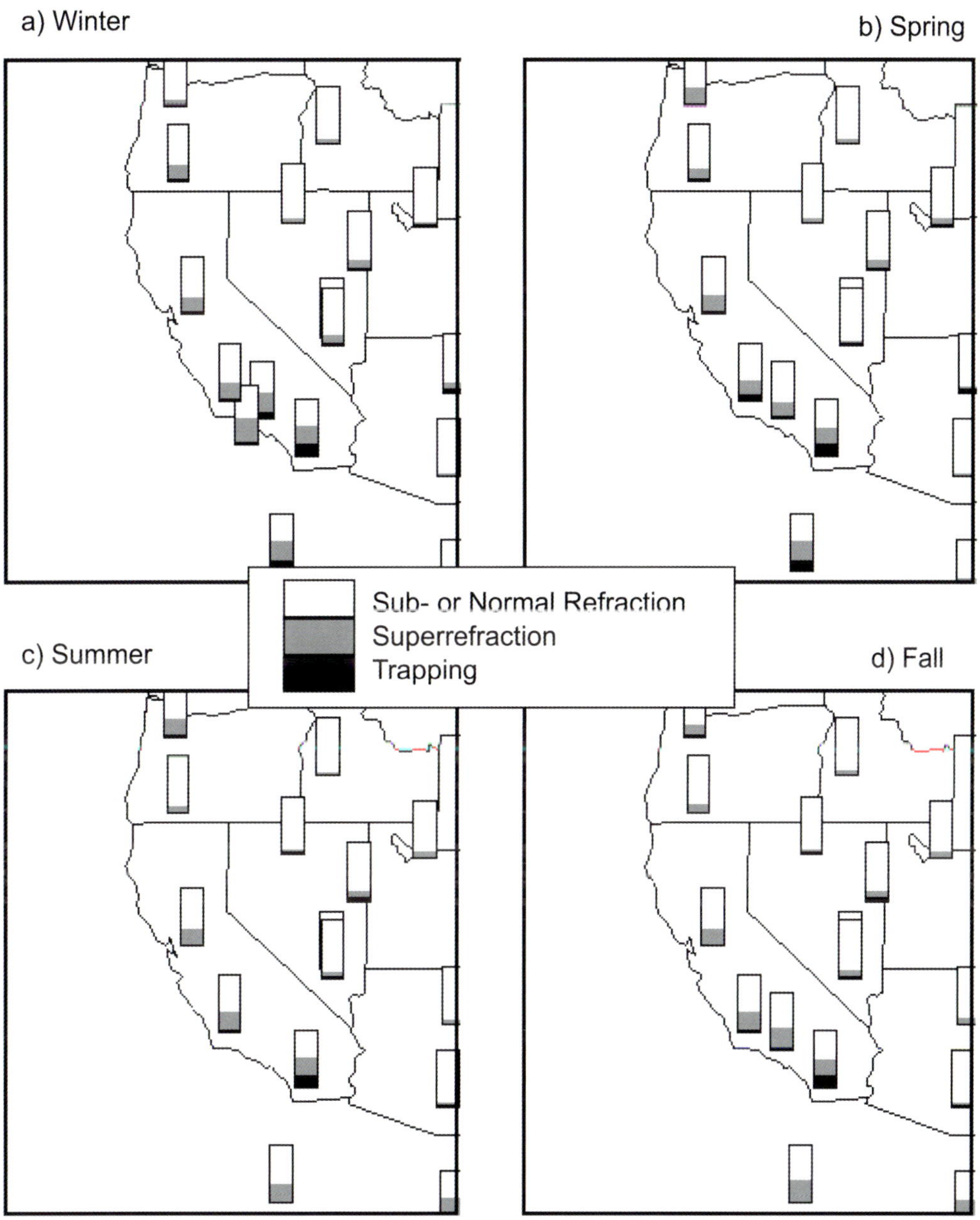

FIGURE 5.4 Same as Figure 5.3, but for soundings taken at 00 UTC. SOURCE: Adapted from Steiner and Smith (2002).

mainly in Southern California essentially throughout the year. The seasonal nighttime patterns depicted in Figure 5.3 are also visible during daytime conditions, as shown in Figure 5.4. Conditions of superrefraction at least once a week are seen throughout the year in the coastal states. In Southern California, superrefractive conditions occur two to three times a week, while trapping conditions exist approximately 20 percent of the time.

Figure 5.5 visualizes the month-to-month variation of the distribution of mean refractivity gradients within 500 m (1640 ft) AGL for soundings taken near San Diego, California. Signal propagation conditions vary relatively little throughout the year. Clearly, conditions of superrefraction and trapping appear to be the dominant modes of signal propagation. This result is in close agreement with temperature inversion statistics compiled based on soundings collected on San Nicolas Island off the shore of Los Angeles by the Naval Air Systems Command Weapons Division (located at Point Mugu) over approximately two decades (Table 5.1). These statistics demonstrate that temperature inversions are a common feature of nearly all atmospheric temperature soundings collected off the shore of California. The average height of the inversion base falls between approximately 300 and 650 m (984 and 2133 ft) with an inversion top at about 500 to 900 m (1640 to 2953 ft), although this varies somewhat diurnally and also by season. Figure 5.6 reveals that during the rainy winter months (i.e., January through April), approximately 80 percent of the inversion top heights fall at or below the altitude of the Sulphur Mountain and Santa Ana Mountain NEXRAD, while approximately 65 percent of the inversion top heights reach higher than the altitude of the Vandenberg Air Force Base (AFB) NEXRAD. Thus, the two radars located at a higher elevation clearly are much less exposed to temperature inversions than the lower Vandenberg AFB radar. Because anomalous propagation conditions appear to be a legitimate basis of concern for radars near the Southern California coastline, minimizing this problem is a valid consideration for siting NEXRADs. In the case of the Sulphur Mountain radar, its higher elevation is necessary for mitigating problems with anomalous propagation.

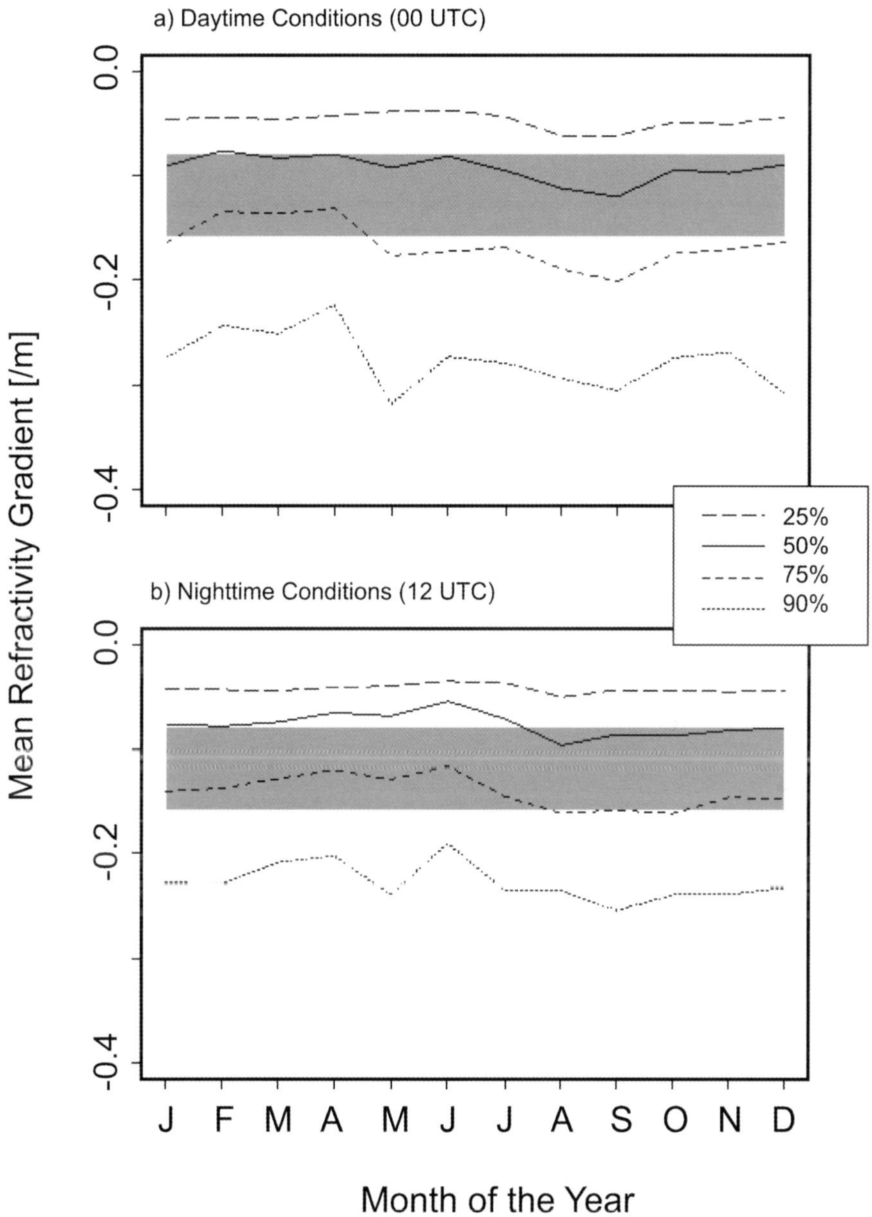

FIGURE 5.5 Distribution of mean refractivity gradients within 500 m AGL based on a 16-year climatology of operational soundings taken near San Diego, California. Shown are contours of equal percentiles of the respective monthly distributions. Superrefractive propagation conditions are indicated by the gray-shaded area. SOURCE: Matthias Steiner, Princeton University.

TABLE 5.1 Temperature Inversion Statistics for San Nicolas Island, California, Compiled by the Naval Air Systems Command Weapons Division Based on Soundings taken from July 1975 through August 1993

	00–08 LST			09–12 LST			13–23 LST		
Month	A	B	C	A	B	C	A	B	C
JAN	94%	331 (554)	521 (579)	88%	474 (528)	693 (520)	90%	501 (556)	718 (576)
FEB	96%	426 (548)	603 (570)	84%	536 (501)	736 (517)	89%	599 (642)	805 (653)
MAR	95%	562 (547)	795 (565)	89%	634 (551)	847 (550)	88%	611 (581)	844 (578)
APR	98%	468 (413)	727 (424)	92%	587 (570)	853 (604)	95%	513 (477)	774 (465)
MAY	100%	487 (333)	799 (371)	100%	550 (408)	877 (439)	99%	481 (336)	796 (377)
JUN	99%	494 (302)	868 (335)	100%	479 (263)	814 (305)	99%	479 (358)	890 (410)
JUL	100%	465 (246)	849 (305)	100%	467 (331)	859 (377)	100%	408 (225)	849 (297)
AUG	99%	476 (258)	828 (322)	100%	416 (271)	757 (300)	99%	473 (277)	845 (313)
SEP	98%	530 (466)	822 (463)	97%	479 (393)	801 (374)	98%	584 (583)	879 (580)
OCT	95%	500 (463)	746 (472)	98%	557 (508)	778 (519)	94%	482 (480)	707 (491)
NOV	98%	394 (467)	614 (508)	91%	466 (446)	691 (464)	94%	453 (474)	707 (499)
DEC	97%	354 (529)	537 (573)	95%	528 (545)	762 (577)	90%	566 (589)	792 (600)

NOTE: The data are grouped into statistics based on soundings collected from midnight to early morning (00–08 LST), midmorning to noon (09–12 LST), and afternoon to midnight (13–23 LST). Column A reflects the percentage of soundings that exhibited a temperature (either surface or elevated) inversion, while column B and C show the mean (standard deviation given in parentheses) height in meters above sea level of the base and top of the inversion layer, respectively.
SOURCE: Adapted from Tables 1, 4, and 7 of Fisk (1994).

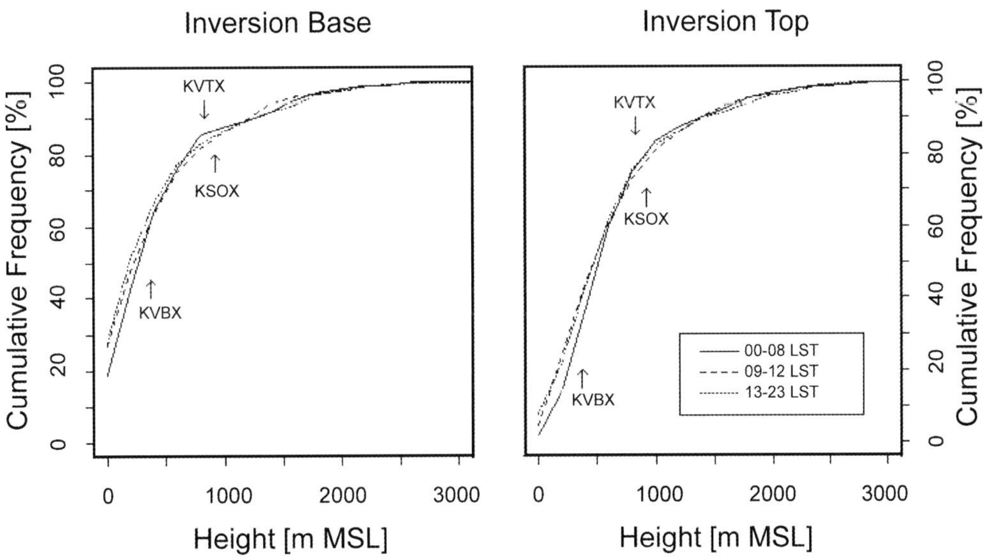

FIGURE 5.6 Cumulative frequency distribution of temperature inversion (a) base and (b) top heights based on soundings collected at San Nicolas Island, California, from July 1975 through August 1993. Shown are statistics for elevated inversions only (i.e., excludes surface-based inversions) for the rainy season months of January through April, stratified by the time of day collected. The data shown correspond to those displayed in Figures 2–7 of the Fisk (1994) report. SOURCE: Matthias Steiner, Princeton University.

6

Flash Flood Warning Process in Los Angeles and Ventura Counties

Forecasting flash floods in Los Angeles and Ventura Counties is a complex process that begins with an understanding of the geographical makeup of the counties and knowledge of the preferential regions of heavy rainfall (i.e., precipitation climatology) for the area of responsibility. The geography and orography of the region for which the National Weather Service (NWS) Los Angeles-Oxnard (LOX) office is responsible are among the more varied in the United States. The elevation ranges from sea level to 10,500 feet over a mere 50 miles, and the land surface varies from ocean to islands, mountains to valleys, desert to agricultural lands to major metropolitan areas. Mountain chains run not only roughly parallel to the coastline but also nearly east-west, making them almost perpendicular to the moist, southerly flow of approaching winter storms.

The relatively flat Ventura County coast is located between the coastal mountain chains of the Santa Ynez Mountains to the north (i.e., near Santa Barbara) and the Santa Monica Mountains to the south (i.e., near Malibu). Two rivers discharge to the Pacific Ocean from Ventura County: the Ventura River's outlet is located just north of the city, and the Santa Clara drains into the ocean just south of the city. Several smaller creeks that originate in the surrounding mountains feed these two rivers, which have a total drainage area of approximately 4500 km^2.

The City of Los Angeles extends mainly southeast of the Sulphur Mountain NEXRAD (KVTX) location. Figures 6.1–6.4 show the main geographic features of the region approximately to the east, southeast, southwest, and west, respectively, as seen from the site of the Sulphur Mountain radar. Figure 6.2 includes a view of the Los Angeles basin and metropolitan area; Los Angeles International Airport is in the direction of 125° from the north. Of the major rivers in Los Angeles County, the Santa Clara River originates in Los Angeles County and then flows into Ventura County, and the Los Angeles and San

FIGURE 6.1 Photograph taken on March 29, 2004, from the site of the Sulphur Mountain NEXRAD looking approximately east-southeast. SOURCE: Julie Demuth, National Research Council.

Gabriel Rivers flow into the ocean. The latter two rivers have a total drainage area of 3056 km^2.

The forecast process requires a forecaster to be cognizant not only of the antecedent hydrological and current meteorological conditions that may trigger a flash flood, but also of modifications of approaching weather systems as they impinge on the complex, mountainous terrain surrounding the highly populous urban areas in southern California. The impact of urbanization on both weather patterns and water runoff and the potential for recent fire-burned regions to modify watershed response and runoff must

FIGURE 6.2 Photograph taken on March 29, 2004, from the site of the Sulphur Mountain NEXRAD looking approximately southeast toward the City of Los Angeles. SOURCE: Julie Demuth, National Research Council.

also be factored into the flash flood warning process. When it becomes clear that the conditions are ripe for the potential for flash flooding, the NWS Los Angeles-Oxnard Weather Forecast Office (WFO) has the primary responsibility for issuing flash flood warnings to the public. The process for communicating and disseminating warnings to the community is complex, involving not only the NWS, but other government and private agencies as well. The end-to-end process of formulating, issuing, and disseminating flash flood forecasts, watches, and warnings to the Los Angeles and Ventura County communities is documented in the following sections.

FIGURE 6.3 Photograph taken on March 29, 2004, from the site of the Sulphur Mountain NEXRAD looking approximately southwest toward the westernmost part of Los Angeles (left) and toward the Pacific Ocean and the Channel Islands (center, right). SOURCE: Julie Demuth, National Research Council.

RAINFALL RECORDS FOR LOS ANGELES AND VENTURA COUNTIES

The rainy season in Southern California occurs from November through March, as shown by the tables of average monthly rainfall (Tables 6.1 and 6.2) and average number of days per month with measurable precipitation (Table 6.3). These tables are based on data from the National Climatic Data Center and the Southwest Regional Climate Center for a sampling of stations. Monthly record precipitation amounts (Table 6.4) and daily record precipitation by month (Table 6.5) indicate that the winter months generally are the ones with heaviest rainfall. However, heavy rains occasionally fall in other

FIGURE 6.4 Photograph taken on March 29, 2004, from the site of the Sulphur Mountain NEXRAD looking approximately west-southwest over the Pacific Ocean and toward the Channel Islands. SOURCE: Julie Demuth, National Research Council.

months, sometimes from thunderstorms or tropical weather systems. Haynes (2001) studied storms that produced at least 3 inches of rain at three or more stations in Southern California and found that most occurred from November through March (Figure 6.5). Table 6.6 illustrates the same information for events of 4 inches or more.

Highest measured precipitation events in Southern California include more than 24 inches in 24 hours (at Haynes Canyon) in mountainous regions. Even with precipitation measured by many rain gauges, actual precipitation measurements still are available for only a small fraction of the overall area of Southern California; thus, it is possible that the largest pre-

TABLE 6.1 Los Angeles Area Average Precipitation (Inches), 1971–2000

	Long Beach	Los Angeles Airport	Los Angeles Civic Center	Santa Barbara
January	2.95	2.98	3.33	3.57
February	3.01	3.11	3.68	4.28
March	2.43	2.40	3.14	3.51
April	0.60	0.63	0.83	0.63
May	0.23	0.24	0.31	0.23
June	0.08	0.08	0.06	0.05
July	0.02	0.03	0.01	0.03
August	0.10	0.14	0.13	0.11
September	0.24	0.26	0.32	0.42
October	0.40	0.36	0.37	0.52
November	1.12	1.13	1.05	1.32
December	1.76	1.79	1.91	2.26
Annual	12.94	13.15	15.14	16.93

SOURCE: The National Climatic Data Center (NCDC).

TABLE 6.2 Other Stations, Variable Lengths of Record

	Riverside	Rancho Cucamonga	Orange	Oxnard
January	2.1	2.4	2.7	2.9
February	2.0	2.2	2.4	2.4
March	1.7	2.1	2.5	2.0
April	0.8	1.1	1.0	0.8
May	0.2	0.4	0.2	0.1
June	0.1	0.1	0.1	<0.1
July	0.1	0.1	<0.1	<0.1
August	0.2	0.2	0.1	0.1
September	0.3	0.4	0.2	0.3
October	0.4	0.5	0.3	0.2
November	0.9	1.5	1.3	1.5
December	1.5	1.9	1.8	1.7
Annual	10.1	12.7	12.6	12.0

SOURCE: Southwest Regional Climate Center.

TABLE 6.3 Los Angeles Area Average Number of Days with Precipitation, 58–67 Years of Data

	Long Beach	Los Angeles Airport	Los Angeles Civic Center	Santa Barbara	Riverside	Rancho Cucamonga
January	6	6	6	6	6	6
February	5	6	6	6	6	5
March	5	6	6	6	6	7
April	3	3	3	2	4	4
May	1	1	1	1	1	2
June	<1	1	1	1	1	1
July	<1	1	<1	1	1	1
August	<1	<1	1	<1	1	1
September	1	1	1	1	1	1
October	2	2	2	2	2	2
November	3	3	3	3	3	4
December	5	5	5	5	5	5
Annual	31	35	35	33	37	39

SOURCE: Southwest Regional Climate Center.

TABLE 6.4 Monthly Record Precipitation Amounts

	Long Beach	Los Angeles Airport	Santa Barbara	Oxnard	Riverside	Orange	Rancho Cucamonga
January	12.8	10.0	13.0	13.2	9.0	12.6	11.2
February	9.4	12.4	13.6	13.8	8.9	11.7	10.5
March	8.8	8.1	11.4	7.6	6.1	10.2	8.1
April	4.4	9.9	5.8	4.2	4.6	5.4	4.8
May	2.3	3.6	1.8	1.0	2.1	1.2	2.5
June	0.9	1.1	0.5	0.6	0.9	1.0	0.7
July	0.2	<0.1	0.9	0.2	1.5	0.3	0.4
August	2.0	0.4	1.9	1.2	2.4	0.8	2.2
September	1.4	5.7	4.1	5.0	3.0	2.2	3.9
October	1.6	2.3	2.4	2.2	2.9	1.7	3.0
November	6.0	6.0	6.9	6.4	5.6	6.7	8.0
December	5.3	8.5	5.9	5.3	6.1	5.5	7.8
Annual	27.7	31.3	40.7	29.9	24.0	31.4	28.8

SOURCE: Southwest Regional Climate Center.

TABLE 6.5 Most Rain Reported in a Day

	Riverside	Rancho Cucamonga	Long Beach	Los Angeles Airport	Los Angeles Civic Center	Santa Barbara	Oxnard
January	3.1	4.0	3.75	4.56	5.71	4.02	5.96
February	2.4	2.3	2.78	3.91	4.25	3.97	4.29
March	2.4	2.2	3.46	3.10	5.87	4.73	4.60
April	1.6	1.6	1.61	1.35	2.74	2.83	2.00
May	1.1	1.5	2.03	1.67	2.02	1.16	1.35
June	0.9	0.5	0.48	0.74	0.76	0.41	0.62
July	1.4	0.3	0.26	0.28	0.13	0.91	0.58
August	1.7	1.9	1.75	2.09	2.06	0.98	0.89
September	2.1	3.3	1.39	1.66	3.95	3.01	3.72
October	1.1	1.6	1.81	1.75	1.71	2.39	1.74
November	2.1	3.8	2.03	5.60	3.85	3.69	3.78
December	2.7	3.9	3.16	2.84	4.85	3.28	5.10
Annual	3.1	4.0	3.75	5.60	5.87	4.73	5.96

SOURCE: Southwest Regional Climate Center.

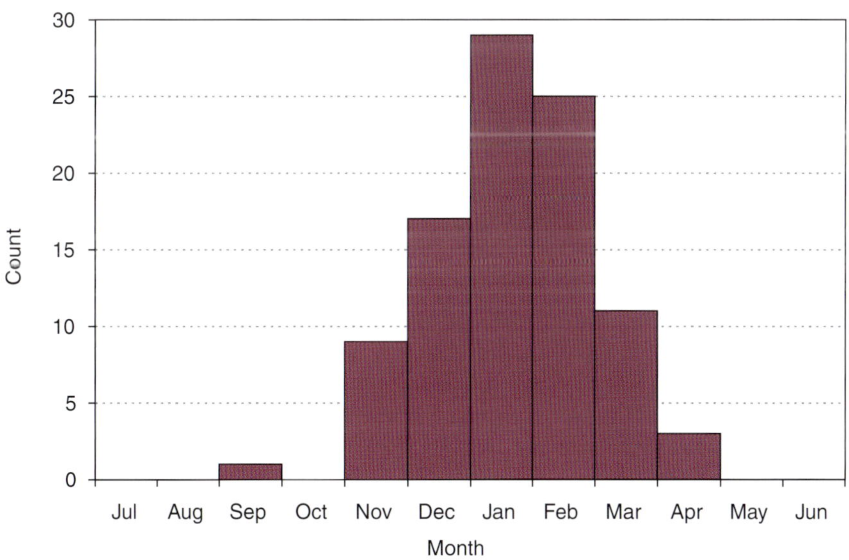

FIGURE 6.5 Monthly distribution of heavy rainfall in which at least three stations in Southern California had at least 3 inches of rain. SOURCE: Haynes (2001).

TABLE 6.6 Days with at Least 4.00 Inches of Rain at One of the Stations Listed in Table 6.5

Date	Location (Rainfall Amount in Inches)
January 17, 1918	Los Angeles Civic Center (4.16)
January 24, 1967	Santa Barbara (4.02)
January 26, 1956	Oxnard (5.96), Los Angeles Civic Center (5.71), LAX (4.56)
February 8, 1962	Oxnard (4.29)
February 18, 1914	Los Angeles Civic Center (4.25)
March 2, 1938	Los Angeles Civic Center (5.87)
March 8, 1968	Oxnard (4.60)
March 15, 2003	Santa Barbara (4.73), Los Angeles Civic Center (4.10)
March 18, 1991	Santa Barbara (4.46)
November 21, 1967	LAX (5.60)
December 5, 1997	Oxnard (5.10)
December 20, 1964	Oxnard (4.44)
December 31, 1933	Los Angeles Civic Center (4.85)

SOURCE: Southwest Regional Climate Center.

cipitation event may not have been measured. In fact, hydrologists (Corrigan et al., 1999) have estimated the probable maximum precipitation[1] over Southern California to be in excess of 30 inches in 24 hours for the foothills and mountainous regions, with the largest value in excess of 48 inches.

THE NWS LOS ANGELES-OXNARD FLASH FLOOD WARNING PROCESS

During the several days prior to the onset of heavy rainfall in the Los Angeles basin, forecasters at the NWS LOX office monitor weather patterns around the hemisphere to detect weather systems that may affect Southern California. They make heavy use of numerical weather prediction model guidance and their own experience to forecast the timing and amount of upcoming precipitation. During this period, they contact emergency management officials (i.e., California Office of Emergency Services) to alert them to the possibility of an upcoming heavy rainstorm. The LOX office may issue flash flood and flood watches to indicate future risks of severe weather to the public. Watches of this type are disseminated on the National Oceanic

[1]The probable maximum precipitation is theoretically the greatest depth of precipitation for a given duration that is physically possible over a given size storm area at a particular geographical location at a certain time of year (AMS, 2000).

& Atmospheric Administration (NOAA) and NWS data circuits and are received by the media and private sector meteorologists for dissemination to the public. The watches are also broadcast on NOAA Weather Radio and displayed on NWS web sites.

Once precipitation is more imminent, the warning process becomes focused on the detection and monitoring of heavy rainfall areas using radar and satellite information and on using these and other observed and forecast data to predict where and when flooding may occur. For the Los Angeles area—where flooding develops quickly both in urban drainage areas and along steep terrain slopes—flash flood warnings are usually issued when the heaviest precipitation bands are offshore to give the public and other users adequate warning times. The NWS Advanced Weather Interactive Processing System (AWIPS) software tracks radar-identified storm cells and projects them forward in a forecast for the next hour.

The main reason the LOX office is forced to issue flash flood warnings an hour or more before heavy rain gets to the potential flood area is that the region is "flashy"—that is, the response time for streams in the area can be as little as 10 minutes from heavy rainfall to flooding. Only at the mouths of the major rivers in the Los Angeles area do response times exceed 2 hours. Thus, in most places, to obtain a flash flood warning lead time of an hour requires that a warning be issued about an hour before the heavy rain begins.

The LOX office obtains flash flood guidance values from the NWS California-Nevada River Forecast Center in Sacramento, California, and from the Ventura County Watershed Protection District (Table 6.7). The LOX office Service Hydrologist provides additional analysis of the situation. These guidance values indicate the rainfall rates and accumulated amounts needed to cause flooding at various locations in the forecast area. The LOX office then combines these guidance values with short-term forecasts of rainfall rate and accumulation to issue warnings of flash floods, urban floods, and slower-moving floods for various locations in the Los Angeles and Ventura County areas. LOX forecasters also monitor rain gauge and stream gauge reports, such as the Automated Local Evaluation in Real Time (ALERT) system (described later in this chapter), and feedback from spotters and emergency managers to indicate how much rain has fallen and where flooding is occurring. There are about 600 spotters trained by LOX forecasters to assist them by reporting severe weather and flooding.

When LOX forecasters determine that a flash flood warning is necessary, a warning is issued indicating the county or counties in which flooding is imminent or in progress, listing the specific locations most at risk. The

TABLE 6.7 An Example of a Hydrologic Conditions Report Issued for Ventura County by Its Watershed Protection District for February 23, 2004

Rainfall Rate or Accumulation	Results
>0.5 in. h^{-1} or 1.0 in per 3 h	Could cause mud and debris flows: 1. Wolf and Piru Burn Areas . . . East Fillmore [below fire area] 2. Along Piru Canyon Rd that could cut off the Lake Piru Recreational Area 3. Simi Fire Area (South Mountain Road, Grimes Canyon Road [Hwy 23 between Moorpark and Fillmore] and Balcom Canyon Road)
>0.6 in. h^{-1} for 30 minutes or longer	Urban flooding in Oxnard near Hwy 101 bridge
>0.6 in. h^{-1}	Urban flooding Ventura RV Park, Meiners Oaks, Casitas Springs, San Antonio Creek
>0.7 in. h^{-1}	1. Can cause flooding along the small barancas within the Simi fire area and can cause localized flooding in Simi Valley and Moorpark. 2. Small area mudslides could occur Santa Monica Mountains (Pacific Coast Highway 23, canyon roads)
>0.7 in. h^{-1} for extended duration	Urban flooding problems in Simi Valley, Thousand Oaks, Moorpark and Camarillo.
>3.0 in. per 6 h at Hopper Mountain (ALERT 662)	Flood problems at 4th Street and Island View Street in Fillmore along Pole Creek.
>4.0 in. per 6 h	1. Flood problems along Hwy 126 at Hopper Creek 2. Some flooding could occur in the Community of Piru. 3. Could fill the debris basins above Piru (Real and Warring Canyons)
>5.0 in. per 6 h	Could cause problems along Tapo Canyon in Simi Valley and at Las Llajas Dam
>6.0 in. per 8 h	Could cause damage at the Simi Sanitation Plant below Simi Valley
>8.0 in.	Problems Hwy 1—Pacific Coast Highway, Local Canyon Roads
>8.0 in. per day	Flooding on the Lower Calleguas Creek and/or Conejo Creek possible
Additional 10–14 in. in the upper watershed	Major problems along the Ventura River and tributaries

warning message is disseminated to NOAA data circuits for distribution to the media and private sector meteorologists and dissemination to the public. The warning is also disseminated to the Emergency Activation System (EAS) network, which is received by the headquarters of the California Office of Emergency Services (CA OES) and then disseminated to its field offices. LOX contacts the CA OES by telephone to ensure that the message has been received. The warning also is broadcast on NOAA Weather Radio. Transmitters are positioned at several locations throughout a NWS office's area of responsibility. Thus, nearly all residents can receive the information if they own the proper receiving equipment, although only a fraction of citizens rely on NOAA Weather Radio for warning information. Law enforcement and emergency management agencies can receive information from this source, but they may also receive it over the EAS or be contacted directly by the LOX office. In dangerous weather situations that have been predicted well in advance, additional NWS staff or community volunteers (e.g., ham radio operators) are sometimes present in the LOX office to assist with communications.

PRIVATE SECTOR ROLE IN FORECASTS AND DISSEMINATION

The NWS flash flood watch and warning statements are disseminated to the public, emergency managers, and law enforcement officials both directly and indirectly (e.g., through the media). The media play a crucial role in disseminating flash flood and urban flood warnings to the public. In addition, meteorologists in the private sector issue advisories and provide other relevant information to the public and other clientele. This group, sometimes referred to as the commercial weather industry (CWI), plays a vital role in conveying weather information to the public and other commercial and government clientele.

CWI meteorologists, sometimes employed by local television stations or operating remotely from a commercial weather service, generally have access to the same weather data and numerical model guidance as do forecasters at the local NWS WFO. CWI meteorologists generally have no access to the exact radar information displayed in the WFO, but they may have alternative and even comparable display systems. Some may have access only to radar data with slightly coarser resolution than in the NWS office and only to a subset of the full suite of scans at various elevation angles. Other CWI meteorologists, however, have their own weather radar and possibly networks of spotters who report rainfall amounts and observations of flooding. Most private sector meteorologists pass along warnings being transmitted by the NWS in addition to their own assessment of the situation. There are nearly two dozen commercial television stations in Southern California, with at least one operating its own weather radar.

Private weather firms play important roles in the hydrological services of the Los Angeles area as well. For example, Fox Weather LLC provides detailed precipitation forecasts to the Ventura County Watershed Protection District. This information is used in water management activities and in anticipation of heavy rainfall events and flooding. Each county in Southern California employs hydrologists, who often assist the NWS by providing flash flood guidance for locations within their counties.

STAKEHOLDERS

The LOX forecasting area includes Los Angeles County, San Luis Obispo County, Santa Barbara County, and Ventura County. Los Angeles and Ventura Counties have two of the larger populations in the United States with approximately 9.8 million and 0.8 million people, respectively. On the smaller end, the populations of San Luis Obispo and Santa Barbara Counties, respectively, are approximately 250,000 and 400,000. Combined, the total population for which LOX is responsible is about 11.3 million people, nearly one-third of California's total population (CADOF, 2002).

Ventura County is home to 10 municipalities—among them, the cities of Ojai, Oxnard, and San Buenaventura (Ventura)—as well as numerous other unincorporated communities, all of which total an area of 4779 km^2 (CADOF, 2002). Within the county's Public Works Agency is the Ventura County Watershed Protection District (VCWPD), which, until January 2003, was known as the Ventura County Flood Control District. The VCWPD exists to ". . . protect life, property, watercourses, watersheds, and public infrastructure from the dangers and damages associated with flood and stormwaters" (VCWPD, 2004). With a population of more than three-quarters of a million people, the county is becoming increasingly urbanized, making the district's task of monitoring watersheds and controlling floodwaters, and LOX's task of flash flood forecasting and warning, increasingly important.

The Ventura County warning system monitors a total of 6825 km^2, which encompasses the Santa Clara River, including Sespe Creek (4175 km^2); Calleguas Creek, including Revolon Slough (834 km^2); and the Ventura River (290 km^2). To monitor this region, the VCWPD relies heavily on the ALERT storm watch system (Box 6.1), which includes data from more than 60 rain gauges and 30 stream gauges in Ventura County (Figure 6.6). An additional 104 precipitation and 47 stream gauges also monitor conditions but do not transmit data in real-time. Compared to Ventura County, Los

BOX 6.1

The Automated Local Evaluation in Real Time (ALERT) System

The ALERT system was developed by the NWS California-Nevada River Forecast Center in the late 1970s (NWS, 1997) with the intention of providing local communities with a low-cost means of acquiring real-time rainfall and streamflow data to better evaluate flood potential. The ALERT system consists of in situ instruments designed to meet a common set of communications criteria. Components of the ALERT system particularly relevant for flash flood forecasting include precipitation and stream gauges (see Figure 6.6). The ALERT precipitation gauge is a standard tipping bucket design. When the buckets tip, a signal is sent to a central ALERT command center, which in turn transmits the information to a flood control district office. Based on the history of precipitation at each ALERT station, an alarm will sound when the amount of precipitation reaches a threshold value that is known to result in flooding. Similarly, when river levels reach a predetermined stage, flood stage sensors send information to the flood control district offices and the local office of the NWS. The NWS issues flash flood watches and warnings accordingly. Real-time data also go directly to the county's Emergency Operations Center and county road crew supervisors to prepare them for the potential of flash flooding.

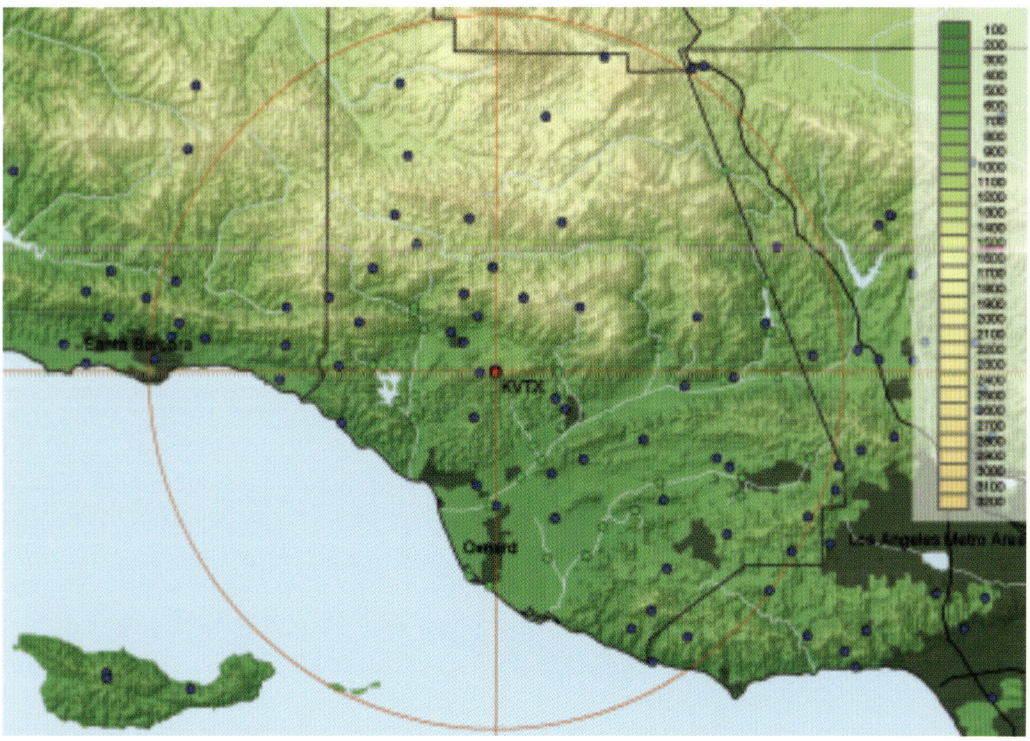

FIGURE 6.6 Topographic map of Ventura County and neighboring Los Angeles and Santa Barbara Counties showing the ALERT rain (blue dots) and stream (green dots) gauges. For reference purposes, the location of the Sulphur Mountain NEXRAD is denoted by the red dot, and its 50-km radius is denoted by the red circle. SOURCE: Image created by Witold Krajewski, University of Iowa. Rain and stream gauge latitude and longitude coordinates from the Ventura County Watershed Protection District.

Angeles County is only twice as large physically—with a total land area of 10,517 km^2—but it houses more than 12 times as many people (CADOF, 2002). The Department of Public Works Watershed Management Division has four main sections dedicated to managing the following major watersheds: Los Angeles River, San Gabriel River, Ballona Creek-Malibu Creek-Dominguez Channel, and Santa Clara River-Antelope Valley. The Los Angeles County Department of Public Works maintains and operates a flood monitoring system consisting of 100 automatic recording tipping bucket rain gauges, 94 of which are on the ALERT system, and nearly 200 nonrecording rain gauges that are observed once daily (LADPW, 2004).

NEED FOR MULTIAGENCY AND PUBLIC COLLABORATION

The NWS is the only agency that has the legal responsibility and authority to issue flash flood warnings. However, a review of the newspaper coverage of flash floods and warnings for Los Angeles and Ventura Counties reveals that numerous local, state, and federal agencies as well as nonprofits and private interest groups are communicating with the public about the risks and appropriate actions to take. A review of the broadcast media would probably also highlight numerous other agencies and parties providing information to the various "publics" in the greater Los Angeles metropolitan area.

While the role of the NWS in producing flash flood forecasts and warnings through the use of NEXRAD and other forecasting tools is essential to the decision-making process for warnings, the public is bombarded with a wide range of information from alternative sources as flash flood situations develop. The watershed districts that operate and maintain the ALERT systems do not officially interact with the public. They are in contact with emergency managers and public officials throughout the region. They use an extensive network of real-time ALERT stream and rain gauge data, a private meteorological service, and their own interpretations of hydrologic models. At the federal level, many agencies are involved and also provide real-time data to enhance the ALERT system. These include the U.S. Geological Survey, the U.S. Forest Service, the Bureau of Land Management, the National Park Service, and the U.S. Army Corps of Engineers. The use of NEXRAD and AWIPS technologies together with forecasting experience allows forecasts and warnings to be made for the counties affected by hazardous weather that develops over the ocean.

The at-risk public must be prepared to take the initiative during a flash flood. Environmental cues, such as intense rainfall for an extended period of time or rainfall when the ground is already saturated, are unofficial warnings.

For example, in January 2004 San Bernadino County—which is outside the LOX WFO's jurisdiction—sent out 113,000 warning letters to residents and property owners advising them that mudslides, debris flows, and flash floods would likely occur with little or no warning with each rainfall for at least the next few years. This action was taken following several bad fires in the area; this "warning" for Southern California may be extended if current dry conditions persist (San Bernardino County Public Information Office, 2004). The decision to alert the public via a letter points out how particularly difficult it is to precisely issue official timely warnings of flooding in areas where watersheds have been significantly altered by extensive burns. One may also wonder what role these messages play in the public's perception of flash flood threats and its willingness to take, or not to take, timely actions when flash floods are imminent.

The warning process is complex and involves the NWS and many others working together to provide timely dissemination of flash flood warnings to the community in a format that both is useful and elicits an appropriate, anticipatory response by the stakeholders and the public at risk.

Recommendation: NWS Weather Forecast Offices nationwide, including the Los Angeles-Oxnard Weather Forecast Office, should continue to expand their collaborative efforts with key stakeholders (e.g., county, police, and emergency management officials) to enhance the effectiveness of flash flood forecasts, watches, and warnings.

The committee commends the Los Angeles-Oxnard WFO on its efforts to communicate with local emergency management officials and the media about potential flash flood situations. However, the committee was informed that the criteria for issuing flash flood warnings are not always sufficient to address all the specific needs of a WFO's end-user community. Existing and new tools—such as geographic information systems (GIS) and the Flash Flood Monitoring and Prediction (FFMP) program, polarimetric radar, and ensemble probabilistic modeling—will make it possible to provide warnings with improved specificity. Moreover, forecasts formulated in a manner analogous to the NWS convective outlooks and with an event severity index analogous to the Fujita scale for tornadoes would provide useful indications of the magnitude of threats. Flash flood forecasts couched in probabilistic terms also would be a useful way to convey uncertainty information. Advanced technologies enable new ways of displaying and communicating individualized warnings to improve the efficiency of the warning process. In the spirit of the recent National Research Council report, *Fair Weather* (NRC, 2003), this should be done in cooperation with the private sector. In addition, in partnership with the media and emergency management officials, the NWS should continue to educate the public about the risks of flash flooding.

7

Evaluation of the Sulphur Mountain Radar and Flash Flood Warnings in Los Angeles and Ventura Counties

As discussed in Chapter 5, the beam visibility of radars sited in complex terrain can suffer from a combination of complete and partial beam blockage, which varies with azimuth depending on the surrounding orography. The visibility of low-level precipitation also is limited by Earth's curvature and, in the case of NEXRAD, by the current restriction to a minimum antenna elevation angle of 0.5°. More fundamentally, the operational effectiveness of a radar also depends on its availability—that is, whether or not it is functioning properly. Both of these aspects have been among the main criticisms of the Sulphur Mountain NEXRAD. For instance, some have questioned the radar's capability to detect low-level weather systems (i.e., below 1.83 km or 6000 ft), including Paris (1997a, 1997b, 1998, 2001), The Rose Institute (1997, 1998), and Fox (2003). Moreover, the General Accounting Office (GAO, 1998) concluded that the Sulphur Mountain radar was not consistently available, increasing the risk that the radar data may be unavailable to the National Weather Service (NWS) and other users when needed for flash flood forecasting and warning purposes. This chapter provides a comprehensive analysis of coverage provided by the Sulphur Mountain NEXRAD and its availability to address these concerns.

RADAR COVERAGE

Sulphur Mountain

Figure 7.1 shows the relief in the 50-km vicinity of the Sulphur Mountain radar location. There are considerably higher mountains to the west, north, and east of the radar site. To the north, the mountainous region includes the Topatopa Mountains with the Hines Peak of 2043 m MSL (above mean sea level) and the Pine Mountain ridge a little further northwest with an eleva-

FIGURE 7.1 The location of the Sulphur Mountain NEXRAD radar (at the crosshairs) and its 50-km vicinity (the red ring). SOURCE: Witold Krajewski, University of Iowa.

tion of about 2000 m MSL. These mountain chains constitute the southern edge of the Los Padres National Forest. Immediately to the east, the Santa Paula Ridge presents the highest peak of 1511 m MSL. To the west, the Santa Ynez Mountains extending in the west-east direction feature several peaks of about 1500 m MSL. To the south, a small South Mountain separates Sulphur Mountain, at an elevation of 829 m, from the coastal Santa Monica Mountains that reach elevations greater than 800 m MSL.

The surrounding orography causes blockage of the beam from the Sulphur Mountain NEXRAD in some directions, especially to the north. A west-east mountain range 10 km to the north of the radar essentially blocks the lower elevation angles over a considerable sector extending from about

300° to 90°, where 0° is to the north. Another small hill immediately to the west partially blocks the radar beam. The radar enjoys an "open view" of the southern sector, which covers directions toward Los Angeles International Airport and the main part of the city and toward the direction of many approaching storms from the Pacific Ocean.

The committee performed detailed calculations of low-level radar coverage and partial beam blockage using high-resolution (i.e., 30-m) digital elevation model (DEM) data and assuming standard atmosphere propagation conditions. These calculations consider the width of the beam and account for the fact that the radar power is concentrated near the beam axis and falls off approximately following a Gaussian function. The calculations consider main beam power loss only, meaning that they do not account for the much lower-intensity sidelobes (i.e., weaker beams that radiate from the antenna in directions away from the main beam). Detailed integration of the power distribution within a beam that may be partially blocked by the terrain allows determination of the percentage of the power loss as a function of distance from the radar. Figure 7.2 shows a conceptual representation of the integration process; for details of the procedure, see Kucera et al. (2004).

The calculations performed by the committee are more realistic than those by Paris (1997a; 1997b; 1998; 2001) in that they account for the entire three-dimensional shape and size of the radar beam; Paris's calculations only considered the location of the main axis of the radar beam. The terrain data used by the committee are also of higher resolution, which has some effect on the results. The committee's calculations do not account for the variable atmospheric conditions that can cause anomalous propagation of the radar beam. However, the committee analyzed a large sample of actual Level II (Crum et al., 1993; Crum et al., 1998) reflectivity data from the Sulphur Mountain radar and concluded that our calculations well represent the essential features of the radar coverage.

The committee performed the calculations for three antenna elevation angles: 0.5°, 0.0°, and –0.5°. According to the current practice of all NEXRADs operated by the NWS, the lowest antenna elevation angle is 0.5°, which forms the "base" scan. The committee's calculations confirm that, at the 0.5° base scan elevation angle, the beam axis rises to 6000 ft MSL at about a 75-km range from the radar. However, half of the beam is still below that level at this range; thus, to infer that there is no coverage below 6000 ft is not correct. The lower half-power (3 dB) point of the beam does not pass above the 6000-ft level until approximately a range of 125 km. Calculations further reveal that using the 0.5° antenna elevation angle, the radar beam pointed in the direction of the Santa Ynez Mountains is only

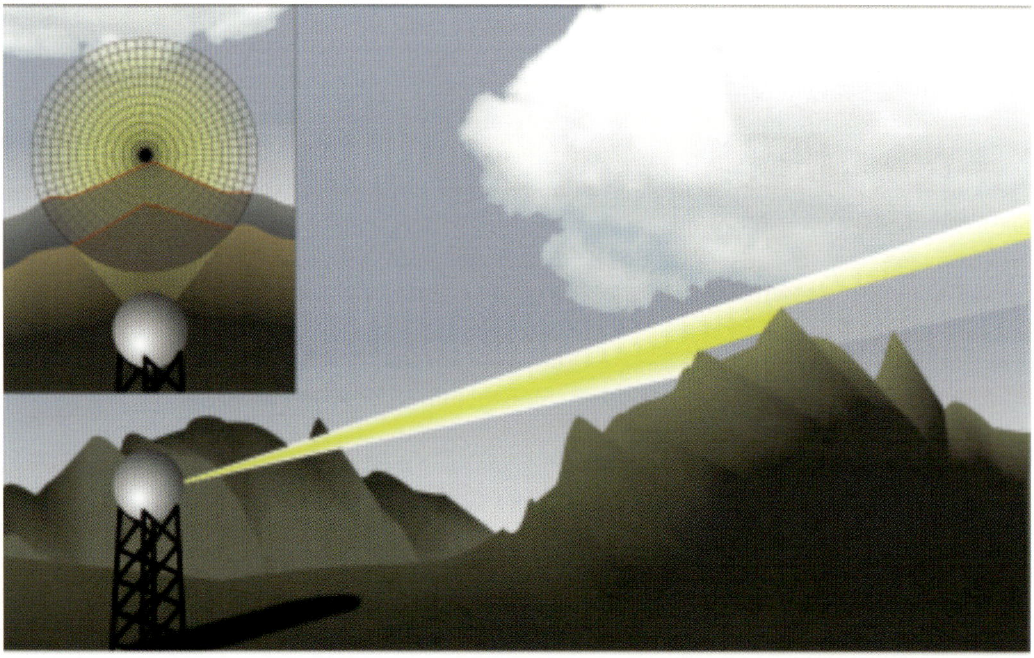

FIGURE 7.2 Schematic representation of the beam power loss integration. The varied intensity of the yellow color signifies the fact that more power is concentrated near the beam axis. SOURCE: Witold Krajewski, University of Iowa.

partially blocked. For example, the beam loses only about one-quarter of the echo power (i.e., assuming a beam-filling target) in the direction of 280° from the north (Figure 7.3[1]).

In the direction of the open ocean at the same antenna elevation angle and at a range of about 100 km from the radar, the beam "illuminates" a portion of the atmosphere that extends from 1.5 km to slightly more than

[1]The mean-sea-level contour of Earth's surface curves down and away from any given point. Under standard conditions of atmospheric refraction, a microwave beam launched horizontally from a point also curves slightly downward with distance. To simplify illustrations such as Figure 7.3, radar specialists commonly depict Earth's surface as flat and show the heights of the beams of interest above that surface. Thus, the radar beams appear to curve upward in the diagrams. The height of the beams is shown correctly, even though neither the surface nor the beams themselves are accurately portrayed.

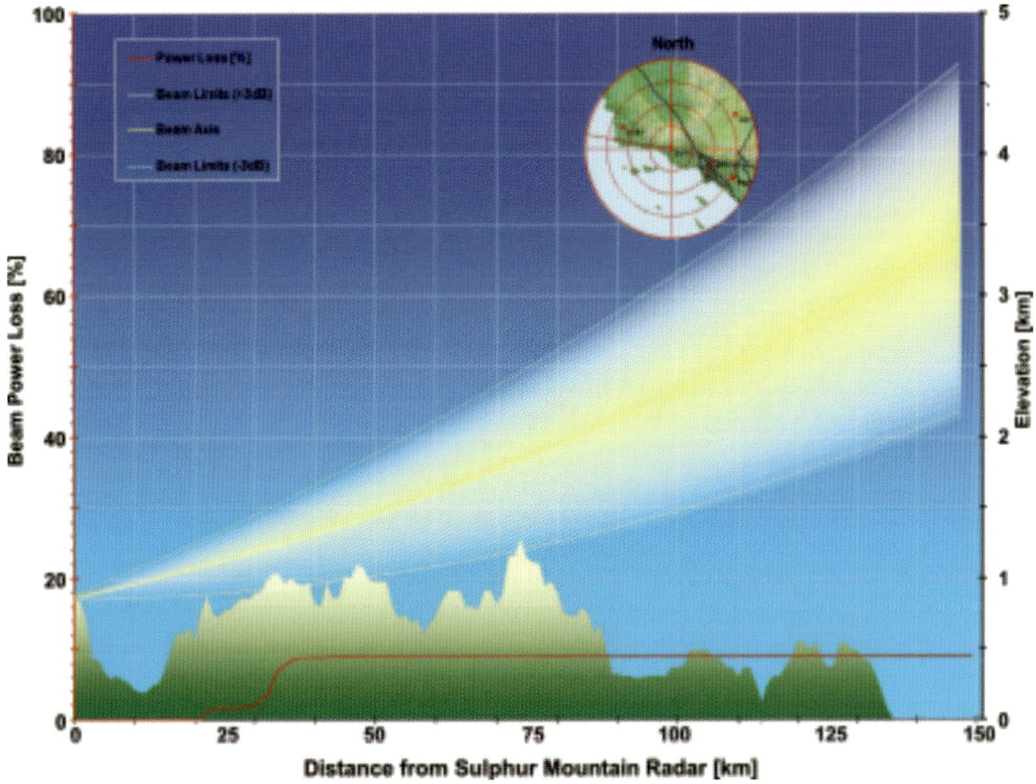

FIGURE 7.3 Sulphur Mountain NEXRAD (KVTX) 0.5° beam and terrain profile in the direction of the Santa Ynez Mountains (280° from the north). SOURCE: Witold Krajewski, University of Iowa.

3 km in the vertical. At the range of 150 km the corresponding portion extends from 2.25 to 4.75 km. Coverage down to the typical altitude of the low-level jet (about 1 km) is restricted to ranges of a few tens of kilometers.

Some researchers (Brown et al., 2002; Wood et al., 2003a, 2003b) have suggested using 0.0° or even negative elevation angles for radars sited at higher elevation, so the committee explored this concept for the Sulphur Mountain radar. By lowering the antenna to the horizontal (0.0°) position, the respective altitude coverage ranges would be 600 m to 2.3 km at a range of 100 km and 1 to 3.5 km at a range of 150 km. This would provide beam-axis coverage at or below 6000-ft altitude all the way out to 125 km as well as some detection capability at the 1-km level out to nearly 150 km. Slightly

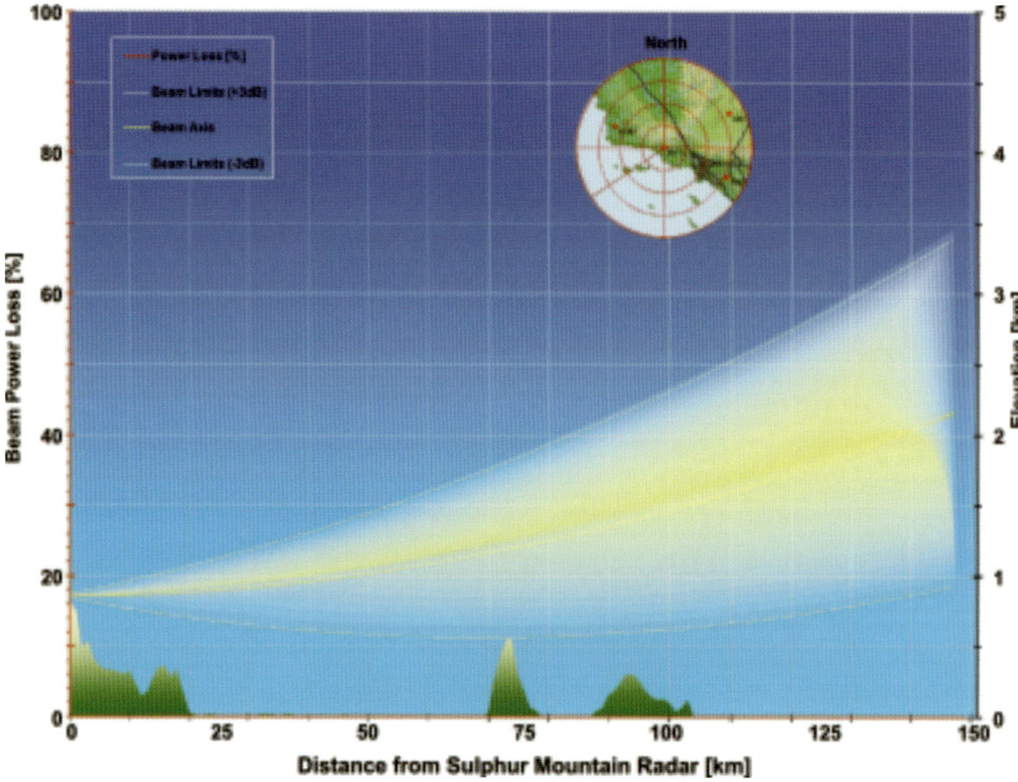

FIGURE 7.4 KVTX radar 0.0° beam and terrain profile in the direction of the Santa Cruz Island of the Channel Islands. SOURCE: Witold Krajewski, University of Iowa.

negative elevation angles could be used to extend the low-level coverage farther out over the water, though with a concomitant increase in beam blockage in some directions. In addition, there would be some increase in side-lobe clutter. Figure 7.4 shows a relevant cross section in the direction of the Santa Cruz Island of the Channel Islands. Lowering the antenna elevation to 0.0° would cause some power loss (i.e., up to about 30 percent) in the beam in the direction of Catalina Island but only over a very narrow sector of about 6° (see Figure 7.5). The direction toward Los Angeles International Airport would remain unaffected (Figure 7.6). Since the radar is situated at an elevated site, the beam would remain well above the ground on its way toward the open waters.

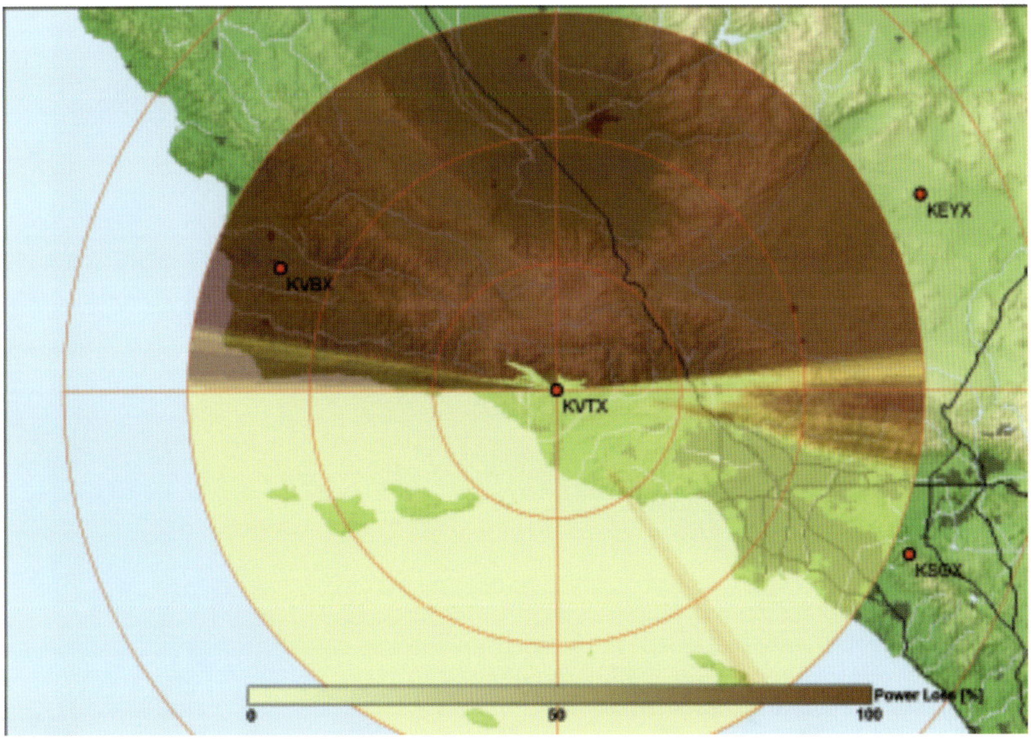

FIGURE 7.5 Power loss distribution of the Sulphur Mountain NEXRAD (KVTX) at a beam elevation angle of 0.0°. Darker colors signify the extent of power loss. SOURCE: Witold Krajewski, University of Iowa.

In addition to Figures 7.3, 7.4, and 7.6, which show cross sections of the radar beam propagation and power loss at just a few azimuths and radar beam elevations, the committee completed calculations at every degree (i.e., 360 azimuths) for the radar beam at 0.5°, 0.0°, and –0.5° elevation. All 1080 of the calculations are available electronically from the following National Academies web site: *http://dels.nas.edu/basc/nexradsm/radar_beam_and_terrain_viewer.html*. This interactive presentation provides the user an opportunity to see how the Sulphur Mountain radar beam propagation and power loss change with direction based on three different beam elevation angles.

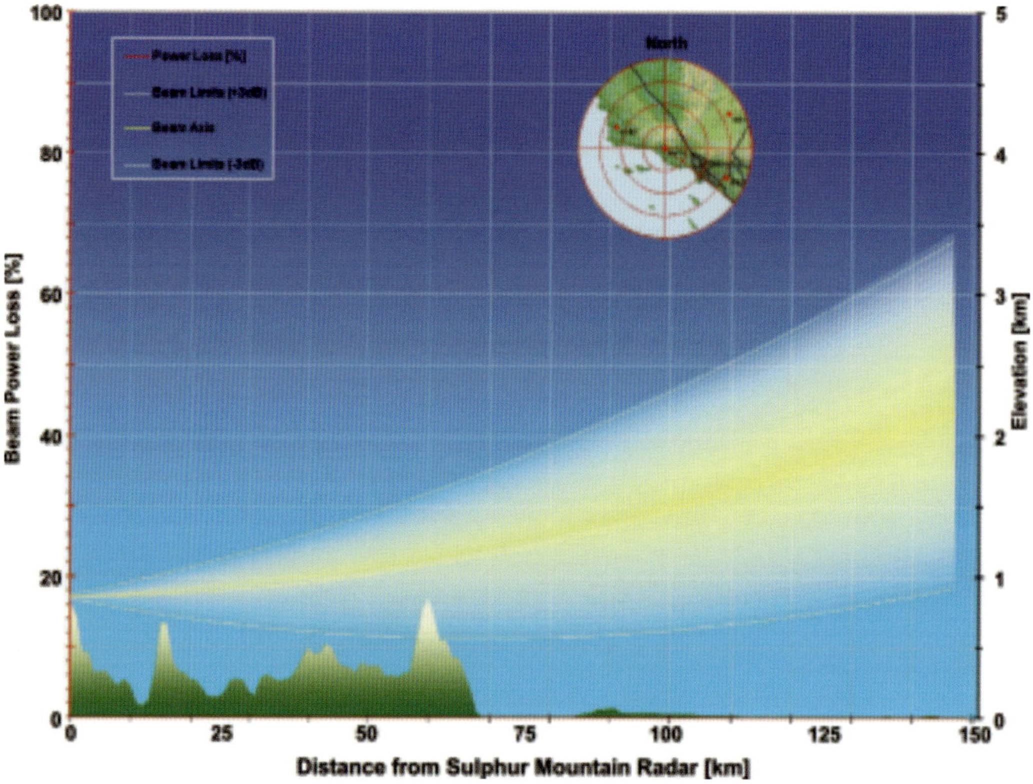

FIGURE 7.6 Cross section toward Los Angeles International Airport for 0.0° elevation angle. SOURCE: Witold Krajewski, University of Iowa.

Adjoining Radars

The detailed calculations performed by the committee for the Sulphur Mountain radar are very labor-intensive and time-consuming; thus, the committee did not repeat them at this level of detail to analyze the coverage from neighboring NEXRADs (i.e., Vandenberg Air Force Base [AFB], KVBX; Santa Ana Mountain, KSOX; San Diego, KNKX; Edwards AFB, KEYX). Instead, the committee performed simplified calculations, similar to those by Paris but using the higher-resolution DEM data and considering the width of the main lobe of the radar beam (i.e., not just the axis). This is a

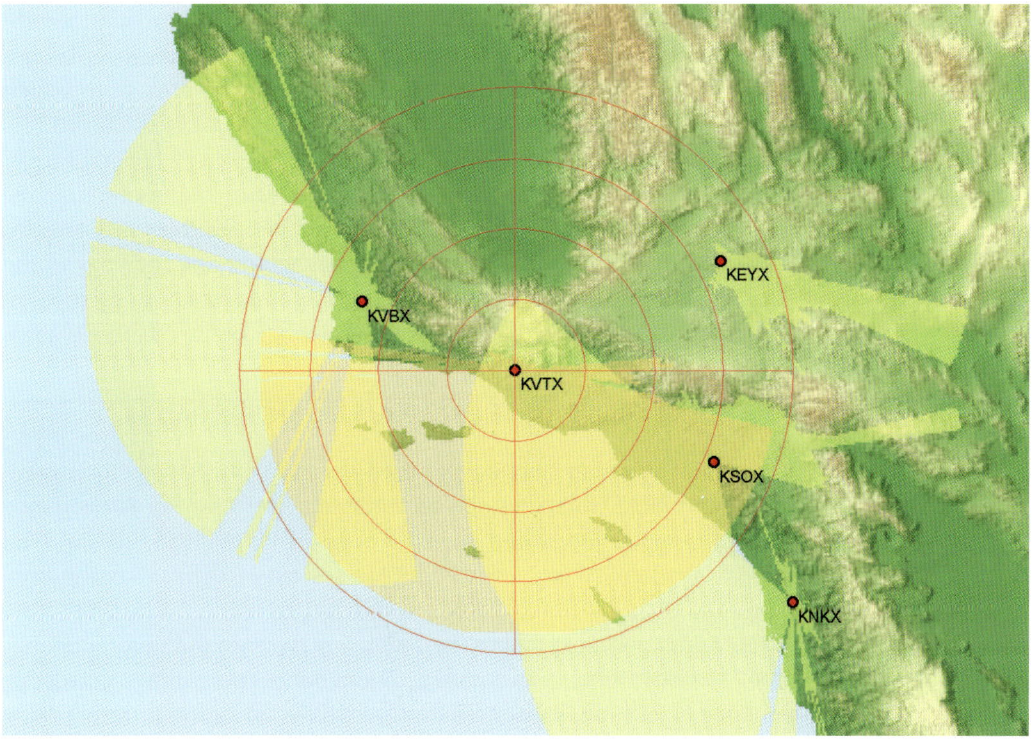

FIGURE 7.7 Southern California radar network coverage of the 0–3 km atmospheric layer. The light brown shaded region shows coverage provided solely by the Sulphur Mountain (KVTX) radar. The coverage shown Is at different elevations, which depend on the distance from the respective radar. SOURCE: Witold Krajewski, University of Iowa.

conservative approach because it reveals only the completely unblocked coverage.

The results of these analyses are illustrated in Figure 7.7, which shows the coverage provided by each respective radar between 0 and 3 km (0 and 9843 ft) MSL. Coverage from the KVBX, KSOX, KNKX, and KEYX radars is shown in light yellow; coverage from the Sulphur Mountain radar is shown in dark yellow; and the area covered by the Sulphur Mountain radar but not by any other adjoining radar is shown in light brown. The latter reveals that the Sulphur Mountain radar provides sole coverage to its southwest, in (1) a north-south swath over some of the Channel Islands, including parts of the

Santa Cruz and Santa Rosa islands and (2) a north-northeast to south-southwest swath west of the Channel Islands. Although these areas are small, they are vital for providing coverage over a key area from which incoming weather systems approach. In addition, diagrams similar to Figure 7.3 would show that KVTX provides better low-level coverage over many areas near the Sulphur Mountain site than do any of the other radars. Figure 7.7 also shows that if the KSOX radar located south of Los Angeles ever fails, the Sulphur Mountain radar provides redundant coverage of the Los Angeles basin. Therefore, the Sulphur Mountain radar is sited in an appropriate position to provide unique, key coverage of the offshore waters.

AVAILABILITY OF THE SULPHUR MOUNTAIN RADAR

A National Weather Service standard exists for the satisfactory availability of all NEXRAD radar systems as outlined in the Weather Surveillance Radar-1988 Doppler Integrated Logistics Support Plan (NWS, 2002a). The operational availability (A_o) of a NEXRAD system is defined as the total operating and standby times (in hours) divided by the total possible time (i.e., 8760 hours per system per year), which is the sum of the total operating time, standby time, preventative maintenance downtime, corrective maintenance downtime, logistics delay down time, and administrative delay downtime. A variation of this measure is the service availability (A_s), which does not include downtime for preventative maintenance and is a more accurate reflection of the radar functionality; it is required to be no lower than 96.2 percent.

When the General Accounting Office (GAO, 1998) analyzed the Sulphur Mountain radar performance, its main criticism was that the radar was not consistently meeting the A_s requirement. Over the 30-month period analyzed by the GAO—from October 1995 through March 1998—the monthly average availability statistics were below 96.2 percent one-third of the time. The availability fell below 90 percent in four of those months, hitting a low of only 79.3 percent in February 1997. In the six months between the period examined by GAO and the time its report was released in October 1998, the monthly averaged availability of the Sulphur Mountain radar fell below the 96.2 percent threshold one other time (Figure 7.8). According to the NWS, the radar was not meeting the availability requirements primarily because of the delay that occurred when a part failed and had to be reordered and installed. The NWS responded to the GAO's analysis by adopting several methods to reduce component failures and to monitor and expedite replacement (GAO, 1998).

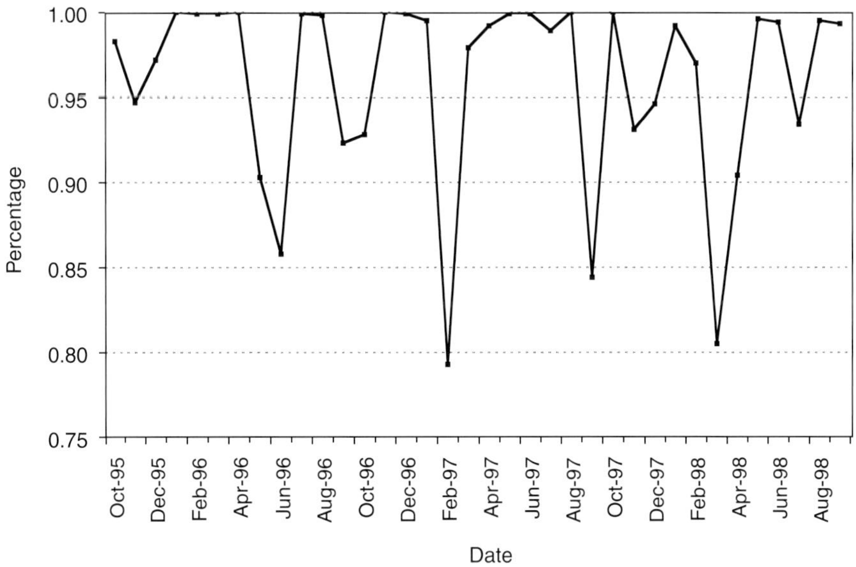

FIGURE 7.8 Monthly average service availability statistics for the Sulphur Mountain radar from October 1995 through September 1998. SOURCE: Compiled by Julie Demuth from data from the GAO (1998) and the National Oceanic and Atmospheric Administration (NOAA).

From the time of the release of the GAO report in October 1998 through December 2003, the monthly average service availability of the Sulphur Mountain radar fell below the 96.2-percent requirement only 4 months out of 63 (Figure 7.9), all of which were due to component failures. The availability of the radar averaged over this entire time period was 98.9 percent, indicating that the Sulphur Mountain radar now exceeds the NWS standard availability requirements and, therefore, is amply functional for use by the Los Angeles-Oxnard (LOX) Weather Forecast Office (WFO) for flash flood forecasting purposes.

FLASH FLOOD WARNING STATISTICS

A directives system exists within the NWS, the purpose of which is to translate the ideas, goals, and principles underlying the NWS mission and strategic goals into action-related items. Among the many areas defined by these directives is one that pertains to operations and services and, more

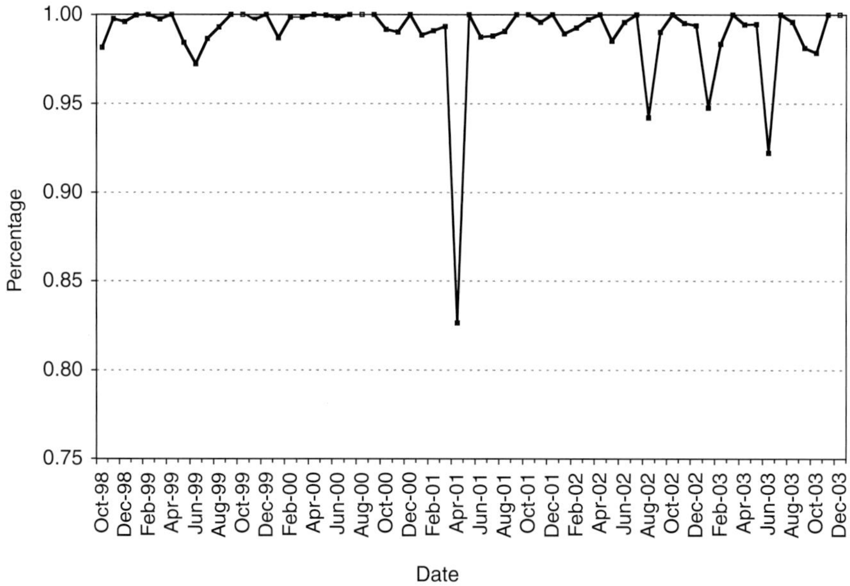

FIGURE 7.9 Monthly average service availability statistics for the Sulphur Mountain radar from October 1998 through December 2003. SOURCE: Compiled by Julie Demuth from NOAA data.

specifically, to performance. The objectives of the performance directive include, among other things, collecting data on significant events and measuring service quality and product accuracy (NWS, 2003b). Accordingly, a directive exists that outlines specific actions to verify various weather watches, warnings, and events, including flash flood warnings (NWS, 2004d). Specific details of the flash flood verification procedures are given in Appendix D.

The NWS uses a number of different measures for verification purposes, from simple, empirical analyses (i.e., number of warnings issued, number of flash flood events) to statistical measures of success (see Table D.2). The latter include the probability of detection (POD[2]) of an event and the false alarm ratio (FAR[3]); a higher POD and a lower FAR indicate better performance.

[2]The probability of detection is the fraction of those occasions on which the forecast event occurred when it was forecast.

[3]The false alarm ratio is the proportion of forecast events that fail to materialize.

State of the Art

To conduct a thorough assessment of the performance of the LOX WFO and the efficacy of the Sulphur Mountain radar, it is important first to establish the "state of the art" by examining the performance of all 116 WFOs in the continental United States. Because installation of the NEXRAD network was completed by the end of 1995, the analysis period used here begins in 1996 and it runs through 2003 to provide a consistent examination of WFO performance using NEXRAD.

The total number of flash flood events that occurred for each WFO during this 8-year period is shown in Figure 7.10. The mode, comprised by 15 WFOs, including LOX, experienced between 25 and 50 flash flood events, for an average of about 3 to 6 events per year. Next most common was to have at least 1 but less than 25 flash flood events, which is the case for 14 WFOs. Of these 14 WFOs, 4 are on the West Coast in orographic regions: Eureka, California (1 flash flood event); Pendleton and Medford, Oregon (20 events each); and Sacramento, California (24 events). On the extreme ends of the spectrum, 3 WFOs recorded zero flash floods—including Seattle, Washington, and Portland, Oregon—and 13 WFOs experienced more than 500 flash flood events within their jurisdictions, including the Austin, Texas, WFO, which experienced 1020 flash floods for an average of more than 125 per year.

The average flash flood warning lead times, which are calculated as the time between the issuance of a flash flood warning and the first occurrence of the event in the warned county, are shown in Figure 7.11 for all WFOs. Once again, LOX, with an average flash flood lead time of 35 minutes, falls in the modal category, which has lead times between 30 and 40 minutes. Of the 116 WFOs, only 22 met the 2004 NWS goal of a 50-minute lead time for flash floods warnings.

During 1996–2003, the mean POD of flash flood events for all stations was 75.5 percent, and the POD for the modal category was between 90 and 95 percent (Figure 7.12). The LOX station is the same as the median value for this statistic, with a POD of 83 percent. Of the four WFOs with a zero probability of detection, three experienced no flash floods during this time and the other experienced only one event.

In comparison with the highly skewed POD distribution, the FAR is more normally distributed (Figure 7.13). Here, the LOX office falls below the median, indicating that it has a better record of not issuing flash flood warnings for events that fail to occur. Of the five WFOs with a FAR greater than 95 percent, four are the cases highlighted in the POD discussion

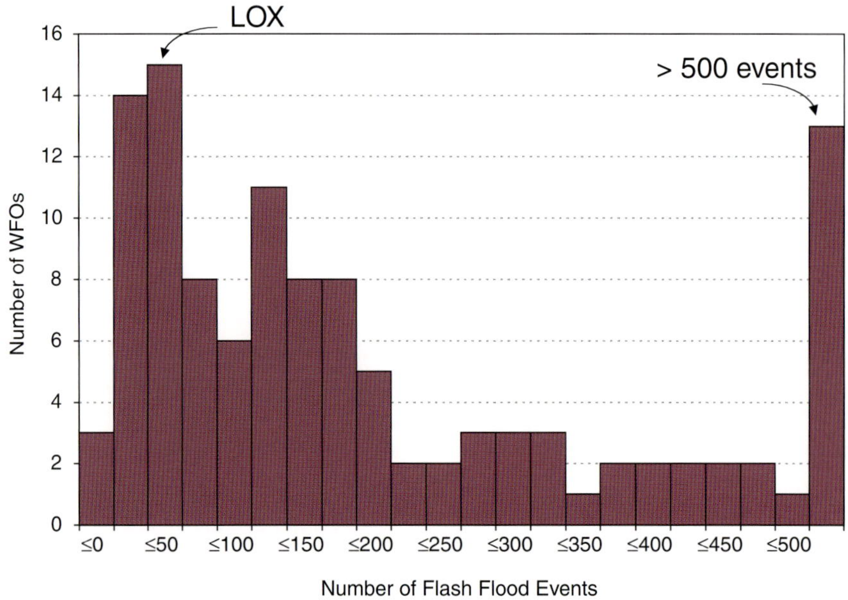

FIGURE 7.10 Histogram of the number of flash flood events occurring for all 116 NWS WFOs in the Continental United States for 1996–2003, after installation of all NEXRADs nationwide. During this time, 47 events occurred within the jurisdiction of the LOX office. The final bar represents the number of WFOs for which between 500 and 1020 flash flood events occurred within their jurisdiction. SOURCE: Compiled by Julie Demuth from NOAA data.

above, three of which had no flash floods and one of which had only one event. In the fifth case, the WFO issued 151 flash flood warnings, only 2 of which were verified. On the other hand, for the six WFOs with a FAR of less than 20 percent, the number of flash floods within their jurisdiction ranged from 91 to more than 500, indicating that their low FAR scores are due to skill and not to a small sample size.

LOX Warning Statistics

The Sulphur Mountain radar, along with the other Southern California radars and other aspects of the NWS modernization of the 1990s, has helped to significantly improve the flash flood warning performance of the

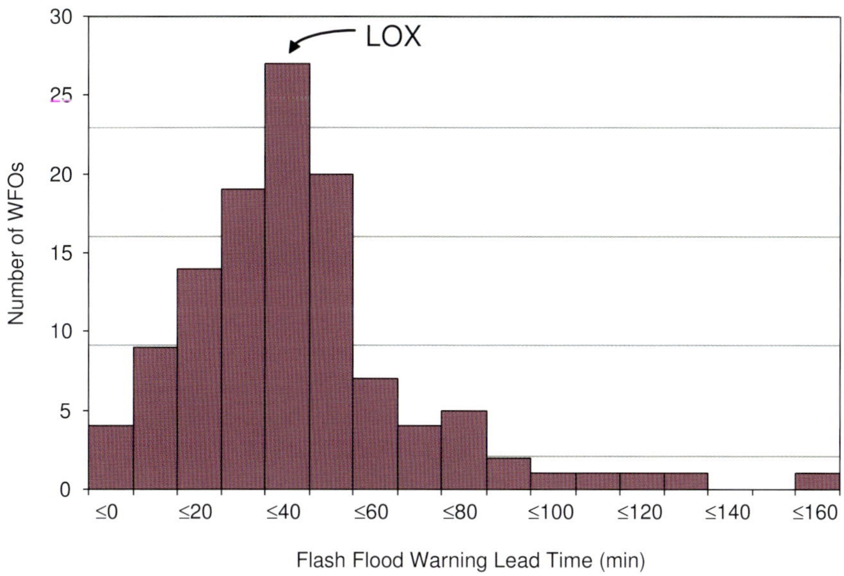

FIGURE 7.11 Histogram of the lead time—the time between the issuance of a flash flood warning and the occurrence of the flash flood—in minutes for all 116 NWS WFOs in the Continental United States averaged over 1996–2003, after installation of all NEXRADs nationwide. The average lead time of the LOX office is 35.0 minutes. SOURCE: Compiled by Julie Demuth from NOAA data.

LOX WFO. Figure 7.14 compares the key flash flood warning statistics for LOX from the pre-NEXRAD period 1986–1994 with those for the 1995–2003 period when NEXRAD was available. All of the statistical measures show improvement, with the POD, the proportion of events with warning lead times greater than zero, and the average lead time more than doubling. Much of this improvement can be attributed to the improved radar coverage available in the NEXRAD era.

The NWS database for 1995–2003 indicates that 15 of the 57 flash flood events that occurred were unwarned. The LOX WFO supplied information indicating that four of the "missed" events were not actually flash floods; one (occurring on January 20, 1997) was not a weather-caused flood, but rather a controlled release into the Los Angeles River channel from upstream reservoirs. Two, listed as occurring on March 5, 2001,

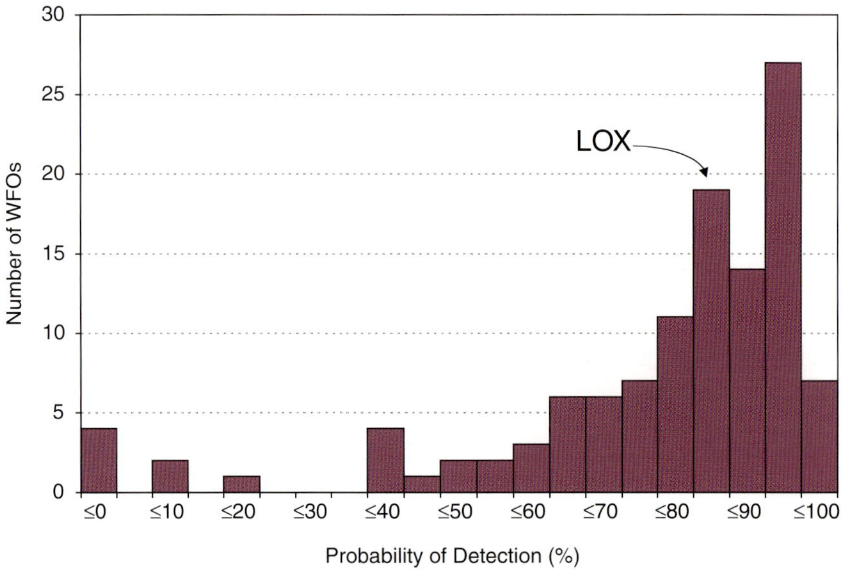

FIGURE 7.12 Histogram of the probability of detection (percentage) of flash floods for all 116 NWS WFOs in the Continental United States averaged over the period 1996–2003, after installation of all NEXRADs nationwide. The POD of the LOX office is 83 percent. SOURCE: Compiled by Julie Demuth from NOAA data.

actually occurred on March 6, 2001, and are being reclassified as floods (i.e., not flash floods) that were covered by flood warnings. Two additional events were covered by warnings not reflected in the NWS database because the early January warnings were issued with the wrong year in the date line. Thus the statistics represented in Figure 7.14 (and Figure 7.15) underrepresent the actual warning performance of the Oxnard office.

Figure 7.15 portrays the key flash flood warning statistics for the LOX WFO, along with national statistics for comparison purposes, for the period 1996 through 2003. The POD and FAR for LOX are similar to the national averages and exceed the 2004 goals for the NWS Western Region which are also shown in the figure. The fraction of events with warning lead times greater than zero is higher than the national average, but the average lead time is shorter. The latter difference may be a consequence of the steeper

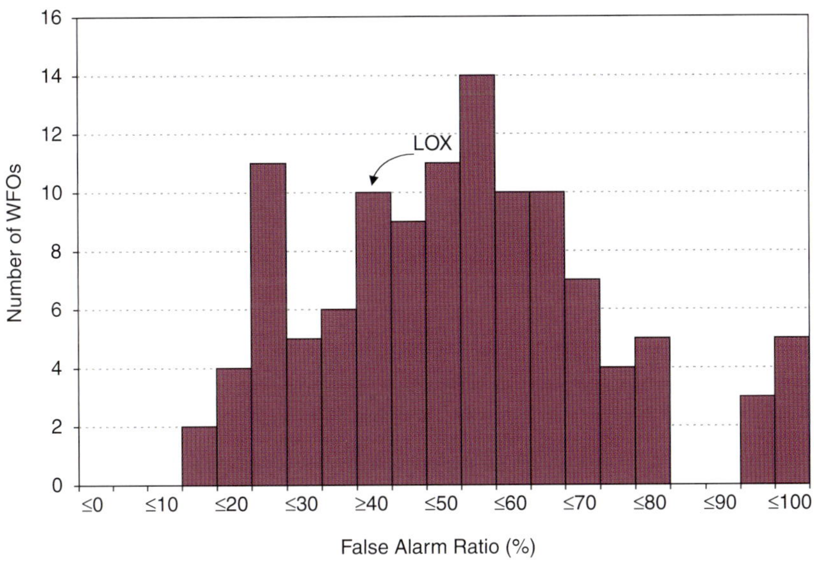

FIGURE 7.13 Histogram of the false alarm ratio (percentage) of flash floods for all 116 NWS WFOs in the Continental United States averaged over 1996–2003, after installation of all NEXRADs nationwide. The FAR of the LOX office is 39 percent. SOURCE: Compiled by Julie Demuth from NOAA data.

terrain and "flashier" streambeds in the county warning areas of concern here, compared to the overall national situation. This is reflected in the relatively short lead-time goal for the Western Region, where steep orography is a major factor.

The NWS database for the LOX WFO for the 1996–2003 shows 47 flash flood events and 136 flash flood warnings, for an average of about 6 events (as noted in the previous section, see Figure 7.10) and 17 warnings per year. This includes events and warnings for all four county warning areas served by the Oxnard office. The WFOs neighboring LOX to the north (San Francisco Bay, MTR) and south (San Diego, SGX) reported 62 and 66 events, respectively, during that period. In comparison to those neighboring California stations, the LOX WFO shows generally better flash flood warning statistics. The San Diego station does have a slightly higher POD, while the San

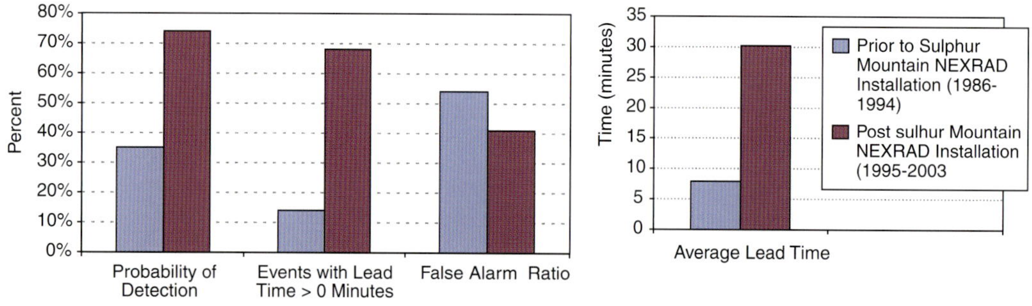

FIGURE 7.14 Comparison of the LOX WFO's statistical performance measures for flash flood warnings from the period prior to installation of the Sulphur Mountain NEXRAD (1986–1994) versus the period after its installation (1995–2003) SOURCE: Compiled by Paul Smith from NOAA data.

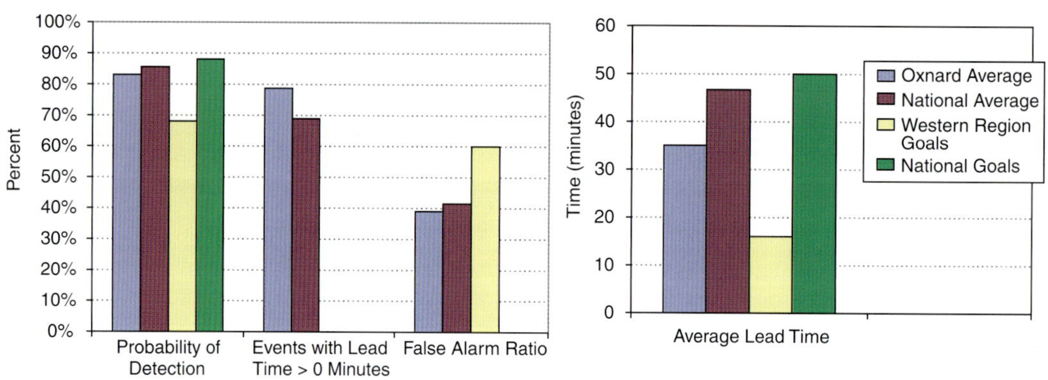

FIGURE 7.15 Flash flood warning statistics for the LOX WFO compared with national statistics from 1996 (i.e., after installation of all NEXRADs nationwide) through 2003. Also shown for comparison sake are the 2004 national and Western Region goals. SOURCE: Compiled by Paul Smith from NOAA data.

Francisco Bay station data indicate an average lead time of nearly 1.5 hours. However, less than half the events in the latter area had warning lead times greater than zero, so the average must reflect numerous cases with lead times of several hours.

The GAO Report

Appendix I of the GAO report (1998) listed flash flood events and warnings for Los Angeles and Ventura Counties for the period 1992 through early 1998. In reviewing that appendix, the committee was unable to reconcile all of the events and warnings listed with the databases maintained by the NWS. Table 7.1 presents a composite chronological listing of events and warnings for those two counties since the Sulphur Mountain NEXRAD was commissioned, as contained in the two NWS databases (i.e., one kept at the LOX WFO and the other used by NWS Headquarters in evaluating the national warning program) and the GAO report. Numerous discrepancies are indicated in the table, with respect to both the numbers and the timing of indicated events and warnings. For example, for the relatively busy period of early 1998 the GAO tabulation indicated 12 flash flood events.[4] These would correspond to 14 events by NWS reckoning, because concurrent events (and warnings) in different county warning areas are tabulated as separate occurrences. Of those 14 events, four have no counterpart in the LOX database and only half are included in the national warning statistics. The main explanation for the latter discrepancy lies in the fact that the NWS generally includes only the first storm event occurring within a given warning period (including overlapping extensions) in its warning statistics. (To include later events would tend to bias the lead-time statistics, because the lead times for the later events would be longer and longer.) The GAO report also omitted one warning-event pair, on February 3, 1998, which appears in the other databases. The time differences present in all but three of the cases result mostly from corrections to the data made after later review of storm reports—sometimes as much as 2 years after the event. For only one of the 1998 events in the table do the entries in all three databases match exactly.

The GAO report not only listed events that do not appear in the NWS databases, but apparently did not take into account the facts that (1) each warning issued by the NWS carries an expiration time; (2) it is common practice to issue updates, modifications, or extensions to the warnings; and (3) consequently, overlapping warning statements were in effect on some of the days. (The warning statements often contain more specific information, intended to be helpful to responders, about threatened areas within the

[4]These are the 12 events in the El Niño year of 1998 mentioned in the attachment to Senator Boxer's letter in Appendix A.

TABLE 7.1 Chronology of Flash Flood Warnings and Events for Los Angeles and Ventura Counties (January 1995–February 1998)

Date	Time (Pacific Standard Time)	LOX Database	NWS Database
4 Jan 1995	0850 (1300)	Warn LA	—
	0930	[a]	FF Ventura
	1000	FF LA (LT 1:10)	FF LA
	1125 (1500)	Warn LA	—
	1250 (1500)	Warn LA	—
	1440 (1800)	Warn LA	—
	1700 (2100)	Warn LA	—
	2000 (0000/5th)	Warn LA	—
7 Jan 1995	1700	—	—
	1735 (2100)	Warn LA	Warn LA
9 Jan 1995	0300	—	—
	2330 (0600/10th)	Warn Ventura	Warn Ventura
10 Jan 1995	0000	—	FF Ventura (LT 0:30)
	0300	FF Ventura (LT 3:30)	—
	0300 (1000)	Warn LA	—
	0400	—	FF LA
	0415 (1000)	Warn Ventura	—
	0600	FF LA (LT 3:00)	—
	0600 (1200)	Warn LA	—
	0717	FF Ventura (LT 3:02)	—
	0800	FF LA (LT 2:00)	—
	0834 (1800)	Warn LA	—
	0834 (1800)	Warn Ventura	—
	0834	—	—
	1305 et seq. (1200/11th)	Warn LA	—
	1305 et seq. (1200/11th)	Warn Ventura	—
24 Jan 1995	2105 (0900/25th)	Warn LA	Warn LA
	2105 (0900/25th)	Warn LA	Warn Ventura
25 Jan 1995	0300 (0900)	Warn Ventura	Warn Ventura
	10 Mar 1995	1850 (0000/11th)	Warn Ventura
	1940 (0400/11th)	Warn LA	—
	2000	—	FF Ventura
	2145	FF Ventura (LT 2:55)	—
	2150	—	—
	2330	FF LA (LT 3:50)	—
20 Jan 1997	1352	[c]	FF LA
2 Sept 1997	1505 (1600)	Warn LA	Warn LA
6 Dec. 1997	0230 (0330)	Warn Ventura	Warn Ventura
	0340 (0445)	Warn LA	Warn LA
	2100 (2300)	Warn LA	Warn LA
3 Feb 1998	0411 (0800)	Warn LA	Warn LA
	0448 (0845)	Warn Ventura	Warn Ventura
	0500	—	FF LA (LT 0:49)
	0500	—	FF Ventura (LT 0:12)
	0600	FF Ventura (LT 1:12)	—

GAO Report	Remarks
Warn LA	Carried erroneous 1994 date
FF Ventura	
FF LA	*b*
—	Extension of 0850 warning
—	Further extension of 0850 warning
—	Extension of 1250 warning
—	Extension of 1440 warning
—	Extension of 1700 warning
FF LA	Not classed as FF in other databases
Warn LA	
FF Ventura	Event actually occurred on 10 Jan
Warn Ventura	
—	May correspond to 0300 event
Listed on 9 Jan	
Warn LA	Carried erroneous 1994 date
—	*b*
Warn Ventura	Extension of 2330/9th warning
FF LA	May correspond to 0400 event above
Warn LA	Extension of 0300 warning
FF Ventura	
—	
—	Extension of 0600 warning
—	Extension of 0415 warning
FF LA	Corresponds to 0800 event
—	Extensions of 0834 warning
—	Extensions of 0834 warning
—	False alarm
—	False alarm
—	Update of 2105 warning
—	Warn Ventura
Warn LA	
—	Corresponds to 2145 FF event?
—	
FF Ventura	Corresponds to 2145 FF event
FF LA	
FF LA	
Warn LA	False alarm
Warn Ventura	False alarm
Warn LA	False alarm
Warn LA	False alarm
—	
Warn Ventura	
—	May correspond to 0745 event
—	May correspond to 0600 event
FF Ventura	

continued

TABLE 7.1 Continued

Date	Time (Pacific Standard Time)	LOX Database	NWS Database
	0745	FF LA (LT 3:34)	—
	0905 (1200)	Warn LA	Warn LA
	0905 (1200)	Warn Ventura	Warn Ventura
	1150 (1500)	Warn LA	Warn LA
	1150 (1500)	Warn Ventura	Warn Ventura
6 Feb 1998	0835 (0945)	Warn LA	Warn LA
	0835 (0945)	Warn Ventura	Warn Ventura
	0910	—	FF Ventura (LT 0:35)
	0910 (1045)	Warn Ventura	Warn Ventura
	0925 (1030)	Warn LA	Warn LA
	0925 (1030)	Warn Ventura	Warn Ventura
	0930	—	—
	0930	—	—
	0955 (1145)	Warn LA	Warn LA
	1000	FF Ventura (LT 1:25)	—
	1010	—	FF LA (LT 0:45?)
	1030	FF Ventura (LT 1:20)	—
	1030 (1200)	Warn LA	Warn LA
	1030 (1200)	Warn Ventura	Warn Ventura
	1038	—	—
	1038	FF Ventura (LT 0:08)	—
	1135	FF LA (LT 3:00)	—
7 Feb 1998	1705 (2300)	Warn Ventura	Warn Ventura
	1800	—	FF Ventura (LT 0:55)
	1950 (0000/6th)	Warn LA	Warn LA
	2040	FF Ventura (LT 3:35)	—
	2205	FF LA (LT 2:15)	FF LA (LT 2:15)
23 Feb 1998	0935 (1800)	Warn Ventura	Warn Ventura
	1115	FF Ventura (LT 1:40)	—
	1130	—	FF Ventura (LT 1:55)
	1240 (1800)	Warn LA	Warn LA
	1240 (1800)	Warn Ventura	Warn Ventura
	1400	FF LA (LT 1:20)	FF LA (LT 1:20)
	1400	FF Ventura (LT 1:20)	—
	1500	—	—
	1500	—	—
	1600 (2000)	Warn LA	Warn LA
	1630	FF LA (LT 0:30)	—
	1800 (2100)	—	Warn LA
	1800 (2100)	—	Warn Ventura
	2050 (2300)	Warn LA	Warn LA
	2255 (0300/24th)	Warn LA	Warn LA
	2255	—	—
24 Feb 1998	0100 (0300)	Warn LA	—
	0300	—	—
	0300 (0600)	Warn LA	Warn LA

GAO Report	Remarks
—	
?	
?	
?	Extension of 0905 warning
?	Extension of 0905 warning
Warn LA	
Warn Ventura	
—	Corresponds to 1000 event?
Warn Ventura	Extension of 0835 warning
—	Extension of 0835 warning
—	Update of 0910 extension
FF LA	Corresponds to 1010 event?
(Two) FF Ventura	Corresponds to 0910, 1000, 1030 events?
Warn LA	Extension of 0925 extension
—	
—	CWA under warning since 0835
—	
Warn LA	Update of 0955 extension
Warn Ventura	Extension of 0925 extension
FF LA	GAO listed wrong county
FF Ventura	CWA under warning since 0910
FF LA	
Warn Ventura	
—	Corresponds to 2040 event?
Warn LA	
FF Ventura	
FF LA	
Warn Ventura	
FF Ventura	
—	Corresponds to 1115 event
Warn LA	
Warn Ventura	Update of 0935 warning
—	
—	
FF LA	Corresponds to 1400 event
FF Ventura	Corresponds to 1400 event
Warn LA	Extension of 1240 warning
FF LA	CWA under warning since 1240
—	Extension of 1600 extension?
—	Extension of 0935 warning?
—	Extension of 1600 extension
Warn LA	Extension of 2050 extension
FF LA	No corresponding event in other databases
—	Extension of 2255/23rd extension
FF LA	No corresponding event in other databases
Warn LA	Extension of 0100 extension

continued

TABLE 7.1 Continued

NOTES: CWA = county warning area; FF = flash flood event in indicated county; LT = lead time from issuance of warning to occurrence of event; Warn = flash flood warning issued for indicated county; times in parentheses in second column indicate expiration time of warning.

[a]LOX WFO contends that this event was ponding of water on farmland due to sustained rainfall, not a flash flood.
[b]LOX WFO provided a copy of a warning actually issued for this event, but because it carried an erroneous date, it does not appear in the official NWS database. In the NWS database, a warning listed for 0500 10 Jan 1994 may be the one related to the 10 Jan event; the warning for 4 Jan does not appear (with either the correct or the erroneous date).
[c]LOX WFO contends that this was not a flood event but a controlled release of water from reservoirs through the Los Angeles River channel.

SOURCE: LOX database information provided by NWS Oxnard WFO; NWS database information provided by NWS Headquarters; and GAO report information from GAO (1998).

designated county warning areas.) For instance, both Los Angeles and Ventura Counties were under continuous flash flood warnings for more than 11 hours on February 23–24, 1998. Nevertheless, the GAO report listed two flash flood events that purportedly occurred at least 10 hours after the first warnings as events having zero lead-time warnings; neither of those events appears in the NWS databases, and the extended warnings were still in effect in any case. The LOX database shows no unwarned events during early 1998 and only two events with warning lead times shorter than an hour. The committee had no means to reconcile all of these discrepancies, but concludes that the table in the GAO report is not a useful indication of the flash flood warning performance of the LOX WFO.

Contributions of Sulphur Mountain and Adjoining Radars

Few flash flooding events in Los Angeles and Ventura Counties have remained unwarned since the Sulphur Mountain NEXRAD was commissioned (see previous section). This radar has been instrumental in the identification of precipitation patterns offshore and tracking them in space and time as they approach and pass over the local coastal mountainous areas. The Sulphur Mountain radar is assisted by two adjacent NEXRADs located at Vandenberg AFB (KVBX) and Santa Ana Mountain (KSOX), respectively. Combined, these three radars provide essential real-time precipitation coverage in support of the work of the LOX WFO.

The primary role of radar is to provide spatial information about precipitation patterns and their intensity and how these change as a function of time. This task may be adversely affected if the radar does not see some precipitation echoes either because of blocking or due to the radar beam overshooting the relevant echoes. Blocking by intervening terrain is almost impossible to avoid completely in complex terrain and may amount to a serious problem. Clever radar siting can go a long way in minimizing terrain blocking effects (see Chapter 4), especially in the directions of primary coverage responsibility. In addition, coverage from multiple radars scanning parts of the same area may further reduce the overall area that is not visible to at least one of them. The area presently covered by the KVBX and KSOX radars—the radars on either side of the Sulphur Mountain (KVTX) radar—is shown in Figure 7.7. By not including coverage provided by the KVTX radar, this figure highlights that each of these three radars covers a unique area and that parts of the domain are illuminated by more than one radar. It is important to recognize that no one of these three radars, on its own, would be able to adequately cover Los Angeles and Ventura Counties.

Low-level coverage is particularly important for quantitative precipitation estimation (see Chapter 5). The current scanning strategy of NEXRAD radars, as noted earlier, limits the lowest scan to 0.5° elevation, which poses a low-level coverage problem for radars mounted at higher altitudes (e.g., on mountaintops). This certainly is an issue for the Sulphur Mountain radar, especially during the winter season when precipitation systems often are shallow (see Chapters 2 and 6). Because the resulting low echo tops only partially fill the lowest-elevation radar beam (i.e., inhomogeneous beam filling), the intensity of such precipitation systems is typically underestimated by the radar. However, forecasters at the LOX WFO are well aware of this and, based on their extensive localized training and experience, are able to compensate for this when assessing the precipitation potential. Moreover, knowing the presence and movement of precipitation systems can be just as important in the flash flood forecasting and warning process as actual estimates of rainfall intensity. The accuracy of radar-based precipitation estimates is also substantially enhanced over land by virtue of the available dense rain gauge network.

Lowering the base scan elevation angle for the Sulphur Mountain radar might yield benefits that go beyond improvements in precipitation pattern identification and tracking (e.g., better observations farther offshore) and quantitative rainfall estimation (e.g., reduced negative effects of partial beam filling and the vertical reflectivity profile). For example, better coverage of low-level winds would result that may reveal local mesoscale flow features (e.g., the barrier jet, which significantly affects the spatial rainfall distribution and amount) and possibly also assist in tornado recognition and associated warnings.

Recommendation: Evaluation of flash flood warnings should be based on their contributions to improved decision making and should employ metrics that take account of the magnitude and scale of the events and the increasing specificity of the warnings. The NWS should improve the database of flash flood events and warnings to include more complete and accurate listings of both warnings and events.

The traditional national measures for evaluating flash flood warnings should be augmented to better take into account the myriad needs of the local population at risk. For example, the value attributed to the lead time of a warning should take into account factors such as the response time of a basin. In addition, warnings indicating the magnitude of the threat and designating specific regions within county warning areas are more useful to both the intended responders and others within the same county who are not affected by the warning. Evaluation statistics such as the probability of detection (POD) and the false alarm ratio (FAR) could be stratified by the scale of the events. The warning and event databases should be adapted to capture the increased specificity of the warning capability, in contrast to the current practice of recording warnings and events at the county level, with little indication of magnitude and impact.

Establishing a reliable database of events for verification purposes involves well-recognized difficulties, but recording the warnings issued is (or should be) a straightforward process. Nevertheless, situations involving a series of overlapping warnings, or extensions of warnings, and possibly multiple flash flood events within a county appear not to be well represented in the national database. This results, at least partly, from the rules employed in establishing that database; such rules may be required to ensure uniformity that permits comparisons across the nation. However, a more comprehensive listing would be useful for detailed analyses.

8

Potential Improvements in Flash Flood Warnings

Advances in flash flood forecasting can arise from a range of improvements that are related to observing capabilities based on radar technology and modeling, data assimilation, and decision support systems. Moreover, the processing power of the Advanced Weather Interactive Processing System (AWIPS; Seguin, 2002) makes it feasible to implement a new Interactive Forecast Preparation System (IFPS; Ruth, 2002), which provides not only standard textual products but also extensive data in digital form (e.g., matrix, graphical, and model interpretation) that is accessible through the National Digital Forecast Database (NDFD; Glahn and Ruth, 2003).

Short-term forecasting of extreme weather relies heavily on the knowledge and experience of individual forecasters and their ability to select relevant information from the real-time flux of observations, followed by interpretation and diagnosis of existing conditions, subsequently leading to inference and decision making (e.g., issuing public advisories and warnings). With the complex system of observations currently in place, it is increasingly difficult for even highly experienced forecasters to process the volume of observations and model output that arrives at their workstations, and to analyze and extract the information necessary to provide effective guidance in real time. This opens a wide range of opportunities for intelligence acquisition technology and the engineering of regional expert systems to capitalize on the human intelligence at local Weather Forecast Offices (WFOs) and to make the most of artificial intelligence applications to process, classify, interpret, and synthesize inhomogeneous data.

Intelligence acquisition technology methods, such as neural networks and support vector machines—for example, working off the NDFD—can be used to extend the lead time of expert decision support system prognostics by tailoring them to the local environment and specific customer needs to increase forecast skills regionally and locally (Kuligowksi and Barros, 1998; Hall et al., 1999; Kim and Barros, 2001; Sivapragasam et al., 2001). Moreover, the implementation of such systems in a probabilistic framework allows direct incorporation of uncertainty in forecast decisions (Murphy 1977; Krzysztofowicz et al., 1993; Applequist et al., 2002).

NEXRAD AND OTHER RADAR SOURCES

Potential improvements to the radar hardware and operation are building on a flexible radar scan strategy and upgrades to polarimetric capability, which will assist in expanded volume coverage, data quality control, and precipitation estimation. The networking of existing information collected by a variety of operational radar sources in addition to NEXRAD will provide redundancy and offers the potential for improved horizontal and vertical coverage, including better coverage at low levels in complex terrain and offshore. Augmentation of the existing operational radar network may be required to extend coverage to important areas not properly covered or temporarily exhibiting an enhanced threat level.

NEXRAD

Scan Strategy

Current NEXRAD scan strategies, and some of the associated limitations, were discussed in previous chapters. Coverage diagrams in Chapter 7 illustrate the low-level coverage provided by the Sulphur Mountain NEXRAD using the current scan strategies; Paris (1997a, 1997b, 1998, 2001) treated the same topic in some detail. The Paris reports considered only the altitude of the beam axis, and because the NEXRAD beam is nearly 1.0° wide, there is some detection capability below that axis (as the diagrams in Chapter 7 demonstrate). Nevertheless, the extent of coverage at and below 6000 feet above mean sea level (MSL) is limited as long as the minimum elevation angle in the scan pattern is restricted to 0.5°.

It is obvious that use of a lower antenna elevation angle from an elevated radar site would provide greater low-level coverage in directions not obscured by intervening terrain. Brown et al. (2002), Wood et al. (2003a, 2003b), and others have noted advantages that could be obtained by using elevations below 0.5°, perhaps even slightly negative elevations, with NEXRADs at mountaintop sites. The discussion in Chapter 7 demonstrates the improvements in low-level coverage that could be achieved by operating the Sulphur Mountain NEXRAD at elevations down to 0°, or at slightly negative elevations.

The NEXRAD antennas are capable of operating at elevation angles down to about –1.0°, and scan strategies using such lower elevations have been proposed. In the original NEXRAD configuration, any effort to implement a change in scan strategy faced a major software undertaking that also would have taxed the data-handling hardware beyond the available capacity. The advent of the open-systems architecture—the open radar product generator (ORPG) already in place and the forthcoming open radar data acquisition (ORDA)—will greatly simplify the process of implementing such changes. The hardware capabilities are enhanced and the software changes needed to implement the new scan strategy become more practical.

The capability to implement NEXRAD scans at a greater variety of elevation angles will soon be in place. However, any change in the NEXRAD scan strategy impacts not only the base data but also the various precipitation products and the many algorithms that operate on those data. The impact of a changed scan strategy on those products and algorithms imposes additional technical hurdles and software requirements; thorough planning and testing will be necessary before implementation of any new scan strategy.

Recommendation: The National Weather Service should improve nationwide NEXRAD coverage of low-level precipitation and wind, especially for elevated radar sites in complex terrain, through the adoption of a modified scan strategy that will allow scanning at lower elevation angles.

The use of lower, and perhaps even negative, elevation angles would allow monitoring of precipitation and wind at lower altitudes and, hence, would provide a more representative assessment of near-surface rainfall rates. Flexible selection of elevation angle steps would allow greater ability to avoid terrain blockage and to capture low-level meteorological phenomena. The NWS should make necessary hardware and software changes to the NEXRAD system to allow this type of modified scan strategy at the Sulphur Mountain site and other NEXRAD installations nationwide.

Precipitation Estimation and Data Quality Control

Dual-Polarization Measurement of Rainfall

Several decades of extensive research have matured polarimetric weather radar technology to a level where dual-polarization radar data can be used operationally to (1) distinguish meteorological from nonmeteorological radar echoes and thereby improve data quality compared to standard Doppler radar; (2) provide a basis for hydrometeor type classification; and (3) improve the measurement accuracy of rainfall over standard Z-R methods (e.g., Zrnic and Ryzhkov, 1999; Doviak et al., 2000; Straka et al., 2000; Bringi and Chandrasekar, 2001; Zeng et al., 2001). The planned upgrade of the NEXRAD network to dual-polarization capability will give great impetus for development of new operational algorithms, in particular, for data quality control and rainfall estimation. There is also increasing momentum for application of dual-polarized weather radars operated at shorter wavelengths (C- and X-bands), and a number of manufacturers are gearing up for, or already offering, dual-polarized Doppler weather radars in their standard product line, as well as dual-polarization upgrade kits for existing radars.

Polarization-diversity radar measurements of precipitation have advanced our understanding of microphysical processes and, through this the estimation of rainfall. The most commonly used polarimetric radar parameters for rainfall estimation are the radar reflectivity factor obtained at horizontal and vertical polarization, the differential reflectivity, and the specific differential propagation phase (Bringi and Chandrasekar, 2001). In addition, the copolar correlation parameter is used extensively for data quality assessment. The differential reflectivity is a good indicator of the average raindrop size within the radar sampling volume and, thus, is used in conjunction with the radar reflectivity factor for improved rainfall rate estimation (Seliga and Bringi, 1976). The specific differential phase exhibits advantages over reflectivity-based parameters because it is a phase measurement that is immune to the absolute calibration of the radar system (e.g., Chandrasekar et al., 1990). Moreover, phase-based parameters are less sensitive to other typical radar problems such as partial beam filling and contamination by hail or terrain (Vivekanandan et al., 1999). At NEXRAD frequencies, use of the specific differential phase for rainfall estimation is particularly valuable for moderate to heavy rain rates. Because of increased level of random errors for phase measurements at lower rain rates, hybrid methods that combine rainfall estimation algorithms based on radar reflectivity, differential reflectivity, and specific differential phase are developed that blend these techniques for the respective rainfall intensity range in which they are best suited (Chandrasekar and Bringi, 2004).

Besides contributing to improved rainfall estimation, dual-polarization measurements are very effective for data quality control, for example, detecting anomalous propagation and ground clutter contamination. Moreover, hydrometeor classification based on dual-polarization radar observations greatly assists in detecting high-density ice particles, such as hail, and the melting layer (bright band), which is useful for vertical profile corrections.

Vertical Profile Correction

Radar precipitation observations are obtained some distance above the ground. The height of the observations above ground increases with increasing distance from the radar due to Earth's curvature. This effect is substantially amplified for radars located on mountaintops. For quantitative rainfall estimation, problems arise because the vertical structure of precipitation intensity varies significantly with altitude. The varying vertical profile of reflectivity introduces uncertainty that can be the dominant source of error in radar precipitation estimates. For example, the radar beam may intersect the melting layer, resulting in a bright band contamination (yielding rainfall overestimation) or illuminate the ice phase of a precipitating cloud system aloft (resulting in a significant underestimation of rainfall). Even within the rainy portion of a storm (i.e., below the 0° level) the intensity may vary with altitude due to evaporation (intensity reduction) or further growth of precipitation particles (intensification).

Taking into account the varying structure of radar reflectivity with height is very important for quantitative radar-based surface rainfall estimation. The precipitation processing system currently in place for NEXRAD (Fulton et al., 1998) does not yet incorporate vertical profile corrections. Several approaches have been developed to deal with this problem, including climatological or short-range profile corrections, range-dependent probability matching, and neural network approaches (e.g., Joss and Pittini, 1991; Joss and Lee, 1995; Amitai, 1999; Vignal et al., 2000; Seo et al., 2000; Liu et al., 2001). To some degree, all involve the vertical extrapolation of observations made aloft to the ground. The NEXRAD ORPG and ORDA developments will ease implementation of such extrapolations, and the resulting improvements in radar precipitation estimates should ultimately enhance NWS flash flood warning capabilities. Such approaches, however, may be less suitable for radars located on a mountain top, where the vertical profile cannot be observed all the way to the ground even at close ranges. Moreover, the complex terrain around such sites may cause significant spatial variations in the vertical profile of reflectivity (e.g., Germann and Joss, 2002) that will be difficult to account for. Coverage of low-level precipitation may thus have to be achieved either by changing the radar scan strategy or by deployment of additional radars at lower altitudes.

Estimation Uncertainty

Characterization of the uncertainty of radar-based rainfall estimates should be as important as the estimates themselves. This realization, however, has not infiltrated the operational radar rainfall estimation, and it is also rarely implemented in research-mode efforts (e.g., Anagnostou et al., 1999; Steiner et al., 1999; Brandes et al., 2001, 2002). Uncertainties in radar rainfall estimates may arise from a number of sources that are related to the measurement accuracy, data quality, retrieval assumptions, and spatial and temporal representativeness of the observations (e.g., Austin, 1987). Measurement uncertainties (e.g., precision, sensor resolution and sensitivity) affect both reflectivity and phase (i.e., velocity) observations, which has implications for the achievable accuracy of radar-estimated parameters (e.g., Metcalf and Ussailis, 1984; Kostinski, 1994). Ground or anomalous propagation clutter, graupel and hail, or melting ice particles, if not properly identified and treated, may significantly increase rainfall estimates (e.g., Balakrishnan and Zrnic, 1990; Steiner and Smith, 2002). Assumptions about the particle size distribution, particle shape, and canting angle[1] determine the retrieved rainfall (Chandrasekar et al., 1988; Jameson, 1989, 1991; Jameson and Kostinski, 1999; Zrnic et al., 2000; Illingworth and Blackman, 2002; Steiner et al., 2004). Moreover, the observations made at some spatial and temporal resolutions may not be representative for surface rainfall without considering the vertical and horizontal structure of precipitation and the associated wind field (e.g., Klazura et al., 1999; Jordan et al., 2003; Morin et al., 2003).

Other Radar Sources

Networking Existing Radars

NEXRAD, particularly those located in regions of complex terrain, are hampered in providing complete radar coverage for the area due to block-

[1]The canting angle is the inclination of the particle minor axis from the vertical.

age of the radar beam by terrain that extends above the radar horizon, as well as by the problem of the radar beam overshooting precipitation echoes below the radar horizon or within blocked sectors. For this reason, it is important that weather forecasters be able to make use of all available operational radar information in order to obtain more complete coverage of the weather within the WFO domain and of weather affecting specific areas of concern. This can be achieved by facilitating access to real-time or near-real-time data from other existing radars within the area, which may include government (e.g., Federal Aviation Administration), military, and commercial (e.g., local TV station) radars.

FAA Radars

The Federal Aviation Administration (FAA) has several networks of surveillance radars installed across the United States to collect radar information in real time and provide this data to the air traffic management units (TMUs) and other aviation end users. Radars deployed by the FAA include the Terminal Doppler Weather Radar (TDWR), the Airport Surveillance Radar (ASR), and the Air Route Surveillance Radar (ARSR). Four of the national FAA surveillance radar networks are shown in Figure 8.1. A brief description of the capabilities of these radars follows, and Weber (2000) provides additional details.

(a) Terminal Doppler Weather Radar

TDWRs are sited near major cities to provide coverage over airports of severe weather hazardous to aviation operations. In particular, TDWRs scan the airspace for detection of low-altitude wind shear, which is known to cause aircraft accidents and human fatalities (Fujita, 1981; Wilson et al., 1984). This 5-cm-wavelength radar measures both reflectivity (weather echoes) and Doppler velocity. It has a 0.5° beamwidth and thus collects higher-resolution data than the NEXRADs. Upgrades to the TDWR Radar Product Generator computer system in 2001 now enable these radars to execute complete 360° volume scans every 3 minutes (2–3 minutes faster than the NEXRADs) while still providing a surface elevation scan every 60 seconds for detection of low-level wind shear. The two lowest radar scans are collected at 0.2° and 0.5° elevation and the typical range of radar coverage is 150 km.

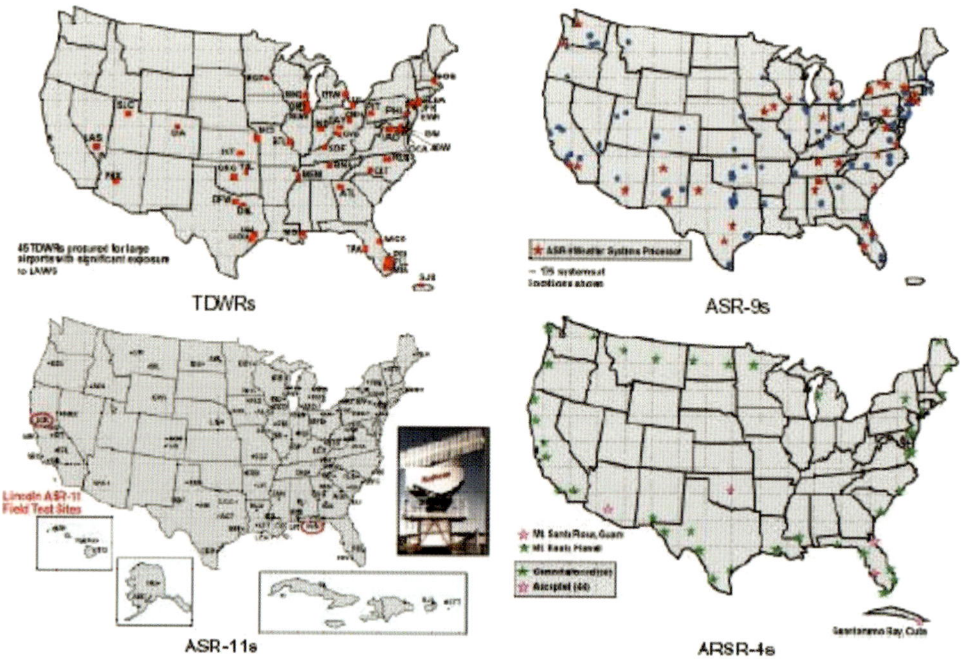

FIGURE 8.1 Locations of FAA surveillance radars. SOURCE: Mark Weber (2000).

(b) Airport Surveillance Radar

The FAA uses ASRs to monitor and track en route aircraft as well as the traffic near major airports. There exist a variety of ASRs (ASR-4 through ASR-11) located at airport facilities throughout the United States. Table 8.1 lists a subset of the ASR facilities located in the southwestern United States. There are twenty-nine ASR facilities operating in California alone, including two at Los Angeles International Airport and one located in Santa Barbara. The radars with the most advanced state-of-the-art capabilities are the ASR-9 and ASR-11 10-cm-wavelength radars. Although these radars operate with a

5.0°-wide, vertically integrated, fan-shaped, elevation beam, they are capable of measuring and displaying radar reflectivity data. ASR-9s scan out to 111 km in range with an update frequency of 1 minute.

In 2001 and 2002, 35 of the ASR-9s across the nation (see red stars in Figure 8.1) were upgraded with new Weather Systems Processors to measure Doppler wind velocity. This includes the two ASR-9s located at the Los Angeles airport. These new radar processors improve detection of weather echoes by eliminating increased ground clutter contamination that occurs under anomalous propagation conditions. The FAA plans to upgrade all of the remaining ASR-9s in the near future with the new radar processor and increase the radar range of all the radars from 111 km to 222 km. Additional ASR-11s are to be installed at smaller airports and Air Force bases across the country.

(c) Air Route Surveillance Radars

The ARSR-4 is a long-range surveillance radar that the FAA has sited along the perimeters of the United States. Two of these radars are located in southern California (Figure 8.1). It is a phased-array system that operates at L-band, produces 10 simultaneous 2° elevation beamwidths, and is able to scan rapidly in azimuth. The ARSR-4 collects reflectivity data, but is limited by a minimum reflectivity level of 30 dBZ and has a problem with properly representing maximum reflectivity within a storm. The FAA and the NWS are currently considering upgrading these radars with a new Doppler weather processor to provide wind velocity information and improved reflectivity data.

Military Radars

The Department of Defense (DoD) requirement for the NEXRAD network was that the radars provide coverage within 65 km for their priority facilities. In certain situations the selected NEXRAD sites were not able to meet the needs of both the military and other funding agencies (i.e., NWS and FAA). As a consequence, several of the military facilities across the United States have their own NEXRADs, such as Vandenberg Air Force Base (AFB) in Southern California. The DoD also operates ASRs at some of its other nonpriority bases. Table 8.1 includes the ASRs located at military facilities in California, including radars at Point Mugu and Edwards AFB.

TABLE 8.1 California (western Pacific region) Airport Surveillance Radar (ASR) facilities, effective 0901Z 24 February 2000 to 0901Z 20 April 2000

City/Airport	HT	ASR	VAR
Bakersfield	37	ASR-4E	14E
Burbank	57	ASR-9	14E
Camp Pendleton		ASR-9	14E
Edwards AFB	17	ASR-5	15E
Fresno		ASR-8	15E
Fresno	17	ASR-4	16E
Garden Grove	37	ASR-9	14E
Long Beach		ASR-8	15E
Long Beach		ASR-9	14E
Los Angeles Intl	17	ASR-9	14E
Los Angeles Intl	37	ASR-9	14E
Marysville AFB	17	ASR-9	17E
Merced		ASR-20	15E
Monterey (Penin)	47	ASR-8	15E
Mountain View		ASR-T	
North Island NAS		ASR-8	13E
Oakland Intl	17	ASR-9	17E
Ontario Intl		ASR-8	14E
Palm Springs		ASR-8	13E
Palm Springs	47	ASR-5	14E
Point Mugu		ASR-7	14E
Riverside		ASR-5	
Sacramento AFB	17	ASR-9	17E
San Diego	67	ASR-9	14E
San Jose (Moffett)	17	ASR-9	17E
Santa Ana	37	ASR-5	15E
Santa Barbara	17	ASR-8	15E
Stockton	17	ASR-7	16E
Stockton		ASR-11	16E

SOURCE: Rita Roberts, National Center for Atmospheric Research.

Commercial Radars

Many television stations across the nation have purchased their own Doppler radar systems and use these data daily in their telecasts. The web site *http://www.weatherexpress.com/wmradar.htm* contains a list and links for local TV station weather radars in the United States. In addition to using radar data in their daily broadcasts, several television networks also provide reflectivity-based, precipitation imagery on their web sites. The KABC television network in Los Angeles operates a 5-cm-wavelength radar that is located north of Northridge, California, on Oden Mountain at an elevation

ARTS	Latitude	Longitude	Ground Elevation
II	35 26 27.9	119 03 35.4	510
III	34 12 14.8	118 21 44.0	740
	33 17 14.1	117 19 47.4	
II	34 52 21.9	117 54 41.2	2330
	36 32 14.4	119 43 04.9	
II	36 46 51.3	119 43 09.7	332
III	33 47 31.1	118 00 08.2	31
	33 49 08.9	118 08 19.8	31
	33 47 31.1	118 00 08.2	
III	33 57 13.4	118 24 28.7	115
III	33 55 56.6	118 24 25.0	115
III	39 07 47.7	121 27 39.7	113
	37 22 34.8	120 33 14.5	
II	36 41 40.9	121 45 28.7	140
	37 25 28.3	122 00 53.4	
	32 42 13.8	117 12 58.8	
III	37 42 22.1	122 13 31.2	3
	34 03 09.0	117 35 39.6	
	33 51 43.0	116 25 49.0	
III	33 50 05.3	116 30 22.7	440
	34 07 06.6	119 07 34.4	
	34 03 15.2	117 35 43.9	
III	38 40 26.7	121 21 55.8	110
III	32 52 59.8	117 08 37.8	450
III	37 25 28.3	122 00 54.4	10
III	33 39 46.1	117 42 47.6	395
II	34 25 26.3	119 50 32.4	65
II	37 53 56.6	121 14 00.6	25
	37 53 15.0	121 14 37.0	

of 3802 feet. This network makes use of its radar and a mosaic of NEXRAD data (including the Sulphur Mountain NEXRAD) in their broadcasts, along with providing routinely updated, full 360° radar information in near real time on its web site.

Data Access and Mosaic Radar Capabilities

Following the installation of NEXRAD across the nation, commercial vendors such as WSI and Kavouras (now Meteorlogix, LLC) and private sector companies such as the Weather Channel have provided the community with national radar mosaics based on NEXRAD-only data. Because of limitations in communication bandwidths, these products have typically used resolution lower than that in the base data in order to maximize processing and display of the combined radar datasets.

Distribution and archival of high-resolution, wideband (Level II) NEXRAD data from more than one radar has been made possible over the past 5 years through the University of Oklahoma project known as the Collaborative Radar Acquisition Field Test (CRAFT; Droegemeier et al., 2002). CRAFT has exploited the existing technology, infrastructure, and low cost of Internet II and the Abilene networks to access and distribute multiple radar datasets to the government, academia, and the private sector. In April 2004, the NWS adopted the CRAFT data distribution methodology, and it has set up three official sites for users to access the 121 NWS NEXRAD radars and some of the DoD radars. The three sites for data access are the University of Oklahoma, Purdue University, and the Education Research Center of the Western Carolinas, with the National Climatic Data Center (NCDC) responsible for archival of all the data via the Internet. The data will also be available via the NWS Telecommunication Operations Center's "Family of Services" approach. Different software packages exist to combine multiple radars in order to provide a comprehensive mosaic of the weather. Figure 8.2 shows a mosaic of weather echoes using a combination of research and NEXRAD data accessed via the CRAFT network.

Based on the success of CRAFT, the FAA funded the Corridor Integrated Weather System (CIWS; Evans et al., 2002) in 2002 to collect and integrate rapid-update TDWR and ASR data together with NEXRAD data to produce real-time maps of radar echoes every 6 minutes over the Northeast corridor of the United States for enroute, air traffic management use. Using this expanded set of radar networks has provided much higher-density coverage of weather over a broad area than has previously been possible.

There is potential to utilize existing, non-NEXRAD radars in a meaningful way to provide expanded weather coverage for WFOs. Access to data from TDWR, ASRs, and research and local TV station radars, especially in regions of complex terrain such as Southern California, can help alleviate gaps in radar coverage resulting from beam blockage by high terrain and problems associated with overshooting beams. Figure 8.3, which does not reflect all of the blockage zones, provides an example of how this may be possible by using an existing ASR-8 located at Santa Barbara to provide radar coverage (blue range rings) in an area that is currently blocked from observation by the Sulphur Mountain radar. Additional coverage along the

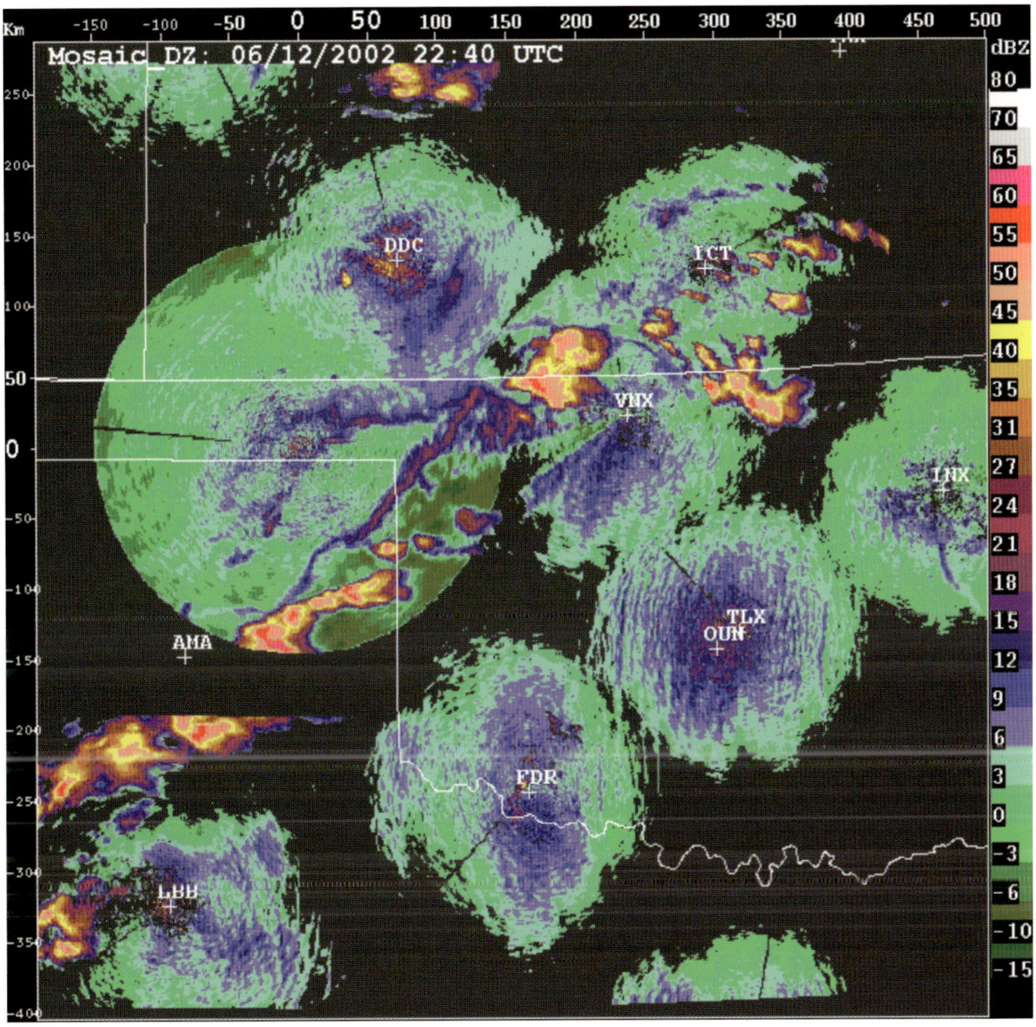

FIGURE 8.2 Mosaic of radar reflectivity data produced from a combination of NWS and DoD NEXRADs and one research radar in Oklahoma, Kansas, and Texas. SOURCE: Rita Roberts, National Center for Atmospheric Research.

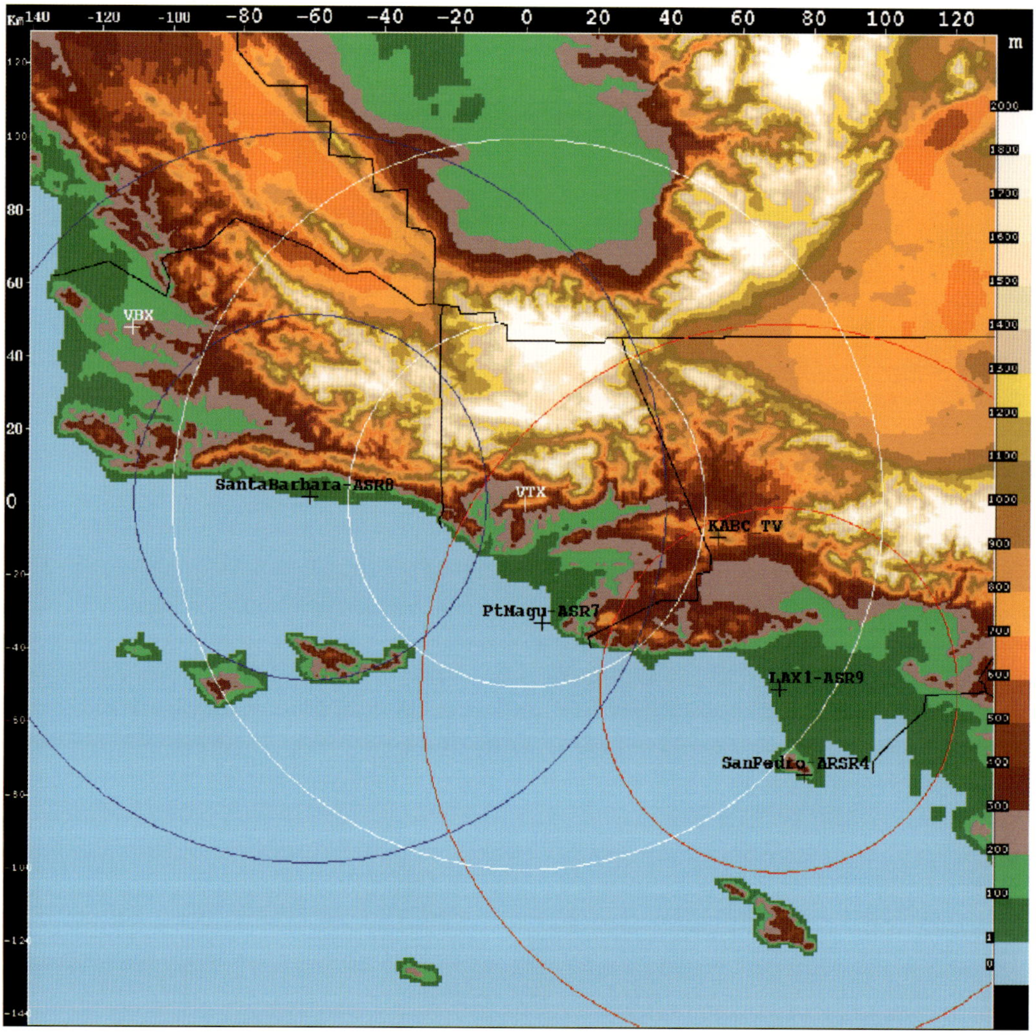

FIGURE 8.3 Image showing how additional radars can augment radar coverage and fill gaps, especially in regions of complex terrain. The white rings (in 50-km intervals) represent radar coverage from the Sulphur Mountain NEXRAD; the blue rings show coverage from an ASR-8 located in Santa Barbara; and the red rings show coverage from an ASR-9 located at Los Angeles International Airport. SOURCE: Rita Roberts, National Center for Atmospheric Research.

coast and over the northern mountain watersheds could be provided by the ASR-9s (red range rings) located at the Los Angeles airport. Access to the KABC TV radar located approximately 15 miles east of the Malibu coast, the ASR-7 at Point Mugu, and the ARSR-4 at San Pedro Hill (see Figure 8.4, which also does not show the blockages) would provide additional radar coverage in the area.

As a further example, in Alabama a cooperative agreement has been instigated among the University of Alabama-Huntsville, National Aeronautics and Space Administration (NASA), the WFO, and the local Huntsville TV network (WHNT) to upgrade the WSR-74C radar to dual-polarization capabilities (Petersen, personal communication). In addition to providing far superior accuracy in rainfall estimation for hydrological purposes, data from this upgraded radar will also be merged with data from other radars in the area to produce mosaics of weather echoes. Through this collaborative effort, all of the above facilities will have real-time access to the upgraded WSR-74C data and surrounding NEXRAD datasets.

Recommendation: To extend radar coverage, all available regional real-time weather radar data should be made accessible to the NWS WFOs, including Federal Aviation Administration (FAA) and Department of Defense (DoD) NEXRAD radars; FAA Terminal Doppler Weather Radars (TDWRs) and other surveillance radars equipped to provide weather-echo data; local television station Doppler radars; and operational radars from other organizations.

High-quality, real-time weather radar data are becoming more widespread in the United States. Although some of these data may not be of the same quality as NEXRAD data, they should be made accessible to NWS forecasters to increase the area and density of coverage of weather radar data—especially in regions of complex terrain—to improve forecasts, watches, and warnings. For radars located along the coast, this could provide a valuable extension of the offshore coverage. The high temporal and spatial resolution of the TDWR data should be of particular value in flash flood forecasting and warning for the major urban areas near which those radars are situated. Data formats and standards must be adhered to and efforts must be made to ensure data quality. This recommendation is consistent with previous National Research Council reports (NRC, 2002, 2003).

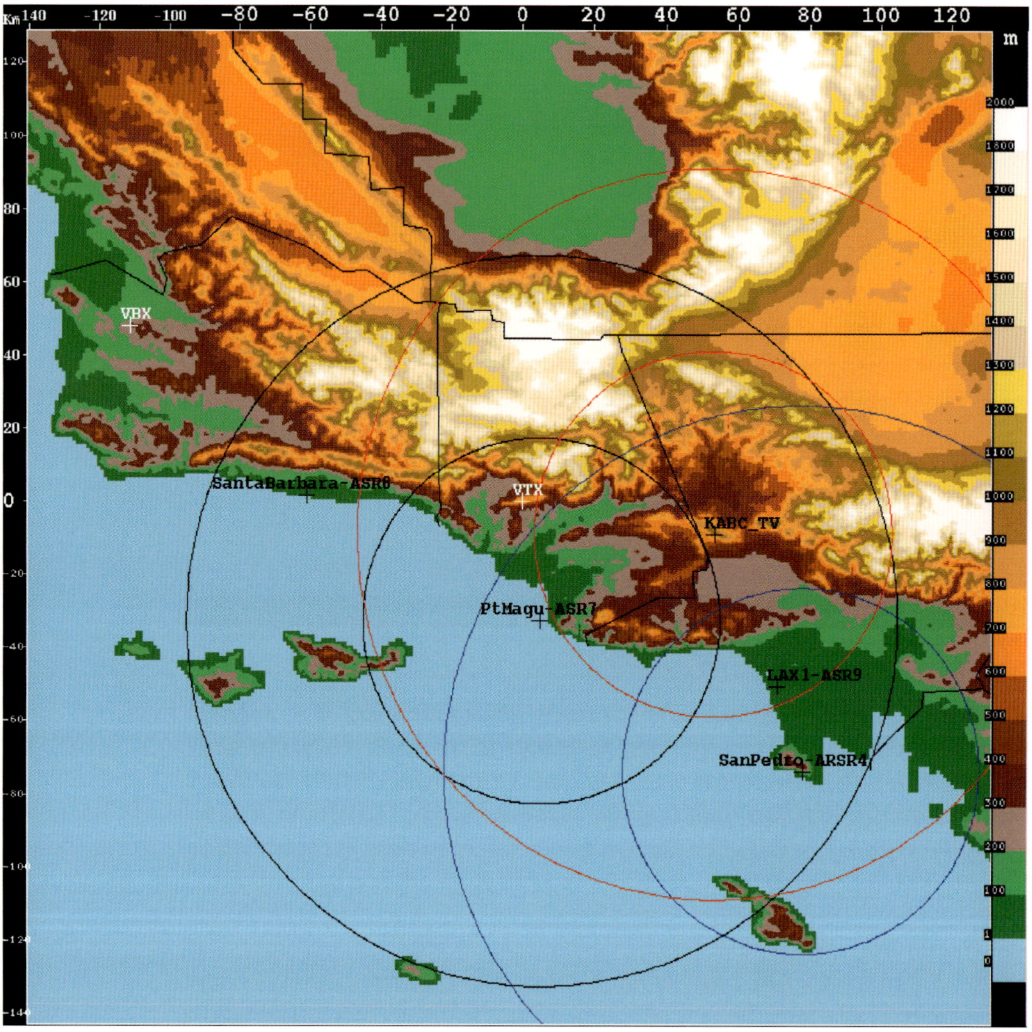

FIGURE 8.4 Similar to Figure 8.3 where the black rings (in 50-km intervals) represent radar coverage from the ASR-7 located at Point Magu along the coast; the red rings show coverage from the KABC-TV 5-cm-wavelength radar located east-southeast of the Sulphur Mountain NEXRAD; and the blue rings show coverage from the ARSR-4 located in San Pedro. SOURCE: Rita Roberts, National Center for Atmospheric Research.

Augmenting Radars

Terrain-induced blockage and limited visibility of low-level weather echoes from mountaintops can significantly reduce the effective radar coverage over a given area. Besides NEXRAD, access to existing, non-NWS (near) real-time radar information may alleviate some of the problems and thus improve the useful radar coverage available to forecasters. Nonetheless, there may still be smaller areas hidden by terrain that would require permanent or temporary (e.g., areas burned by wildfires) radar coverage. Deployment of additional radars may solve this problem (Figures 8.5 and 8.6). Because only short-range coverage typically is needed, these additional radars may operate at shorter wavelengths and thus be smaller, cheaper, and possibly mobile (Bluestein et al., 2001).

A drawback of radar operation at higher frequencies than the NEXRAD S-band (10-cm wavelength) is the precipitation-induced signal attenuation. Often it is hard to separate intrinsic reflectivity changes from changes caused by attenuation from intervening precipitation. At X-band frequencies (3-cm wavelength), the precipitation-induced attenuation can be as much as 20 dB or higher. The amount of attenuation degradation increases with range from the radar site; however, by operating such radars over short distances only, this effect is minimized. In addition, recent advances in dual-polarization radar technology have resolved this problem fairly well in rainfall (e.g., Bringi and Chandrasekar, 2001; Matrosov et al., 2002; Anagnostou et al., 2004).

Figure 8.5 shows a schematic of the role that could be played by smaller, short-range coverage radars. The transition from single, large radars

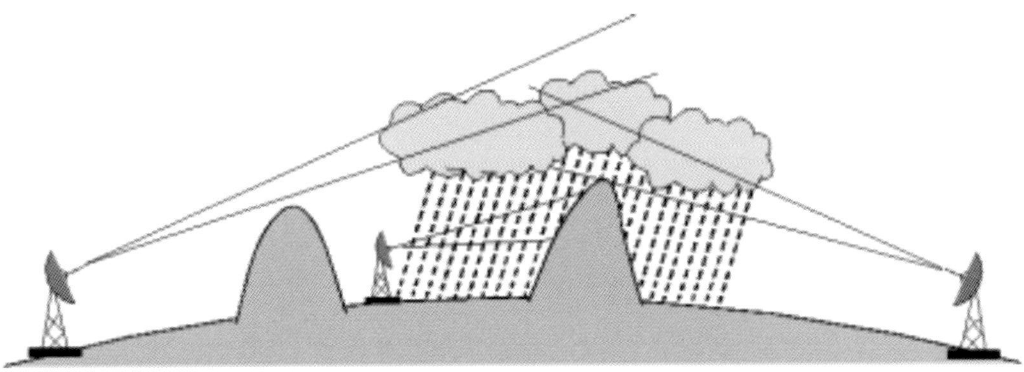

FIGURE 8.5 Schematic of short-range radars and the coverage they provide in orographic regions. SOURCE: V. Chandrasekar, Colorado State University.

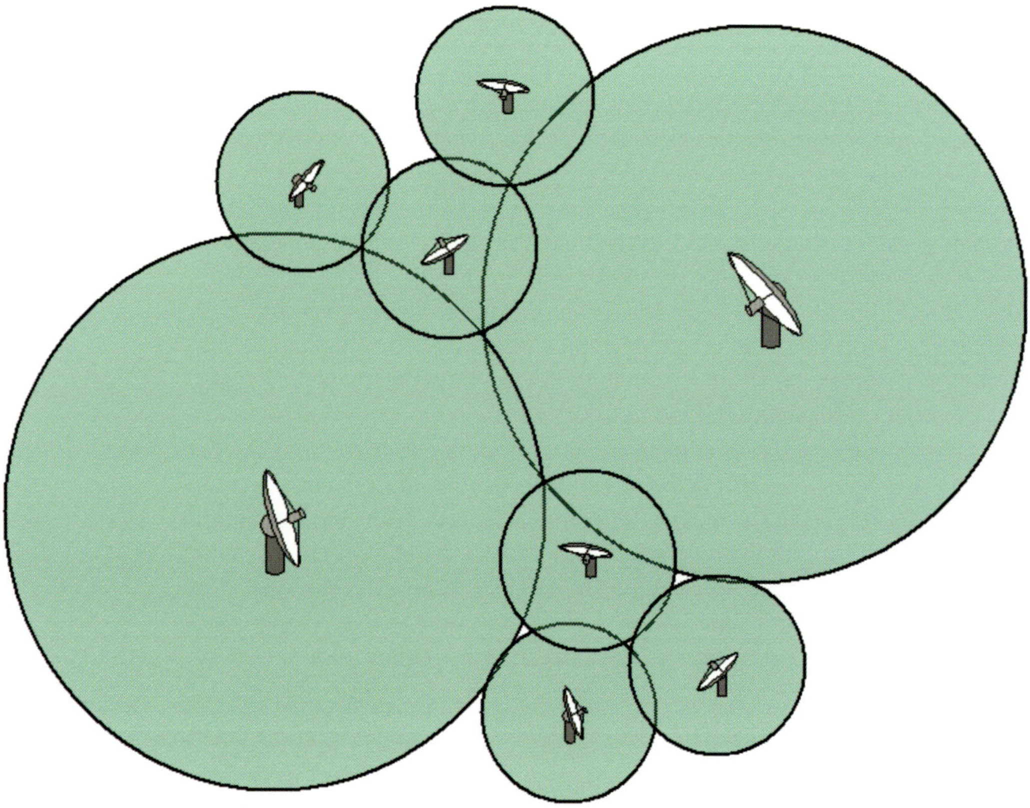

FIGURE 8.6 Schematic of a network of smaller radars that operate as a distributed, collaborative adaptive system. SOURCE: V. Chandrasekar, Colorado State University.

that provide long-range coverage to a set of smaller, cheaper radars is to be accompanied by building the architecture and protocols for these systems to operate together so that data from all of these smaller radars can be integrated. The degree of flexibility increases with many radars, and a distributed, collaborative adaptive strategy can be implemented onto it. The network of small radars can also be deployed adaptively, depending on the application. Figure 8.6 shows a schematic of a network of large and small radars that operate as a distributed, collaborative adaptive system. The NWS currently is collaborating with academic institutions that are developing this technology.

Recommendation: The NWS should consider augmenting the NEXRAD network with additional short-range radars to improve observation of low-level meteorological phenomena.

The NEXRAD is designed for long-range coverage with a single-beam antenna that provides coverage of a large volume of the atmosphere. Although this is adequate for many situations, NEXRAD coverage at low altitudes far away from the radar can be insufficient due to Earth's curvature and terrain-induced blockage. Additional temporary local problems may arise, for example, as in areas affected by recent wildfires. Some of these problems can be resolved by deploying additional radars, which could be smaller, cheaper, and more easily (and even adaptively) deployable than NEXRAD. To maximize their use, these systems should be networked together, data formats standardized, and metadata established.

MODELING, DATA ASSIMILATION, AND DECISION SUPPORT SYSTEMS

National numerical weather prediction (NWP) models, such as the National Centers for Environmental Prediction (NCEP) ETA model, provide forecasts of precipitation rate and accumulation for the contiguous United States, and forecasters have access to these models daily. Regional-scale nonhydrostatic mesoscale NWP models may be initialized and nested within global- and national-scale models to provide higher-resolution predictions at the regional to local scale (Doyle, 1997; Mass et al., 2002; Grimit and Mass, 2002). Given the vast improvements in computing resources, the use of these models is no longer restricted to a research mode; they now can be run on workstations at WFOs in support of the forecasters (e.g., Manobianco et al., 1996).

Quantitative characterization of specific states of the atmosphere and their uncertainty as a result of both model and observation errors is key to data assimilation in NWP models. In contrast, detection, classification, and space-time tracking of weather features places different requirements on data assimilation in the context of operational forecasting, which has led to the development of expert decision systems.

Modeling and Data Assimilation

Mesoscale Modeling and Data Assimilation

A relatively new weather prediction model, called the Weather Research and Forecasting (WRF) model (Michalakes et al., 2001), has been developed collaboratively by several agencies: the National Center for Atmospheric Research (NCAR), NCEP, Forecast Systems Laboratory (FSL), Center for Analysis and Prediction of Storms (CAPS), Air Force Weather Agency (AFWA), and Naval Research Laboratory (NRL). The WRF infrastructure has been designed to provide more flexibility and operational efficiency in the ability to nest mesoscale models within the national grid, to select the domain size and grid spacing used, or to run ensemble forecasts, and flexibility in the use of cumulus parameterization schemes or the explicit treatment of convection on high-resolution (e.g., 4-km) grids. Additional improvements provided by this model can be found at: *http://www.wrf-model.org.* National-scale precipitation rate and accumulation forecast products are available for viewing on-line at this web site daily, and forecasters at several WFOs across the United States have taken advantage of the opportunity to access and use these products.

One of the most promising attributes of the WRF model is the improved prediction of mesoscale convective systems (MCSs) using explicit treatment of convection on a 4-km grid, compared to prior forecasts of MCSs that employed a cumulus parameterization scheme on a 10-km grid (Done et al., 2004). The WRF forecasts with explicit convection were found to be far superior to the 10-km-grid forecasts in their ability to forecast the MCSs that actually occurred, the number of MCSs that would occur each day, and the organization of the precipitating systems. There also is evidence that the explicit version of the model can predict finer distinctions in the precipitation structure, with features such as bow-echoes, lines of supercell storms, and the convective and stratiform regions of squall lines, than possible with some of the existing operational models. Plans are under way to run and assess the performance of the WRF model during the 2004–2005 winter season (Weisman, personal communication). It is anticipated that improved prediction of the structure, organization, and intensity of precipitating storms in the forecast models will lead to improved 12- to 24-hour outlooks for severe weather, including the potential for flash floods.

Assimilation of Doppler radar data with high spatial and temporal resolution has been shown to provide improved accuracy in the representation of three-dimensional wind, thermodynamic, and microphysical fields in mesoscale and cloud-scale numerical models (Verlinde and Cotton, 1993; Sun and Crook, 1997, 1998, Xu et al., 2001; Weygandt et al., 2002a, 2002b; Snyder and Zhang, 2003; Sun, 2004; Xiao et al., 2004; Zhang et al., 2004) and to improve short-term forecast skill over existing numerical models. One such system that has been run operationally at the Washington, D.C.-Baltimore, Maryland, WFO is the four-dimensional Variational Doppler Radar Analysis System (VDRAS; Sun and Crook, 2001) which assimilates single-Doppler radar data into a numerical cloud model. VDRAS provides three-dimensional wind fields used by forecasters to analyze wind shifts and surface convergence in the atmospheric boundary layer (0–3 km altitude) that are precursors to the development of convective precipitating storms (Wilson and Schreiber, 1986). A different data assimilation scheme that uses Newtonian nudging has been employed by Xu et al. (2003) to forecast winds and precipitation accumulation associated with winter storms. This system ingests NEXRAD data from several radars in the northeastern United States into a mesoscale numerical model (MM5) and is run operationally for the FAA.

Techniques such as those mentioned above, which produce high-resolution wind fields that are updated every 6–12 minutes, could be a significant asset for NWS forecasters in regions such as the California coast, where wind speed and direction play a critical role in the distribution and intensity of rainfall. White et al. (2003) cited the importance of monitoring the orientation of wind speed and direction in the lowest few kilometers of the atmosphere for the onset of low-level jets that trigger upslope flow and enhanced rainfall (exceeding flash flood guidance thresholds) along the coastal mountains of Southern California.

Coupled Modeling

Coupled hydrological-meteorological operational systems have been developed and run for the purpose of integrating local precipitation prediction models with hydrological models on the catchment scale and for generating short-term flash-flood warnings with hourly forecasts (Georgakakos and Hudlow 1984; Georgakakos, 1986b, 1986c, 1987; Bae et al., 1995; Miller and Kim, 1996; Westrick and Mass, 2001; and Westrick et al., 2002). The greatest benefit achieved in using a coupled approach was found to be when the forecast lead time was comparable to the response time of a flash flood-prone catchment of interest (Georgakakos and Foufoula-Georgiou, 1991). Current approaches in the United States and Europe, particularly where water management and flash floods in complex terrain are of uppermost concern, now incorporate the use of weather radar data to provide quantitative precipitation estimation (QPE) and numerical model-based quantitative precipitation forecasts (QPF) for hydrometeorological forecasting systems (French and Krajewski, 1994; Lee and Georgakakos, 1996; Dolcine et al., 1998; Yates et al., 2000; Johnson et al., 2001; Mecklenburg et al., 2002; Cress et al., 2002). One such system currently being tested operationally is a prototype software system that provides operational estimation and prediction of rainfall and streamflow in the mountainous subcatchments areas of the Panama Canal watershed in Panama, Central America (Georgakakos, 2002). This fully automated system incorporates information from radar reflectivity and hourly rainfall data, discharge and surface weather reports from Automated Local Evaluation in Real Time (ALERT) rain gauges, upper-air radiosonde observations, and ETA model forecasts. Use of the system fosters a close collaboration between meteorologists and hydrologists on a day-to-day basis. Verification of the short-term predictions of rainfall and streamflow has been very good. Another notable example of an operational coupled hydrometeorological-hydrologic forecast system is employed in the Pacific Northwest (Westrick and Mass, 2001; Westrick et al., 2002). The accuracy of the observation-tuned MM5-DHSVM (fifth generation Pennsylvania State University-NCAR Mesoscale Model-University of Washington Distributed-Hydrology-Soil-Vegetation Model) simulated peak streamflows exceeded that achieved by the NWS Northwest River Forecast Center (RFC) results, although the results were highly sensitive to data density and quality.[2]

Prediction Uncertainty

Numerically predicted meteorological parameter fields, such as temperature, wind, and precipitation, exhibit uncertainty that depends on the model physics, initial conditions, and forecast lead time (e.g., Cortinas and Stensrud, 1995; Harris et al., 2001; Germann and Zawadzki, 2004; Walser et al., 2004). Estimation of the prediction uncertainty can be achieved through probabilistic ensemble modeling based on perturbed initial conditions or simulations based on varied physical parameterizations (e.g., Du et

[2]The National Research Council recently released a report assessing the USGS National Streamflow Information Program (NRC, 2004). The conclusions and recommendations from this report may improve the quality of streamflow data, which in turn could lead to improvement in the development and calibration of hydrologic models.

al., 1997; Stensrud et al., 1999; Richardson, 2000; Krishnamurti et al., 2000; Tracton and Du, 2001; Mullen and Buizza, 2001; Ebert, 2001; Toth et al., 2003; Grimit and Mass, 2002). The range of model results provides a measure of the anticipated agreement or disagreement of the various forecasts, with smaller ranges, suggesting higher confidence in a particular forecast.

Similarly, the uncertainty of hydrologic predictions based on stand-alone hydrologic models that use radar-estimated rainfall as input (e.g., Ogden et al., 2000; Sharif et al., 2004) or coupled meteorological-hydrologic models that may also use numerically predicted rainfall can be estimated by probabilistic means (e.g., Ganguly and Bras, 2003; Nijssen and Lettenmaier, 2004). Seo et al. (2003) present the latest NWS developments for real-time assimilation of hydrometeorological and hydrologic data into operational hydrologic forecasting. A special issue of the *Journal of Hydrology* (2004) is devoted to the Distributed Model Intercomparison Project, a NWS-spearheaded effort to evaluate coupled hydrometeorological-hydrologic models and assess associated uncertainties in streamflow prediction.

Decision Support Systems

Success in short-term forecasting of extreme weather has heretofore relied on the knowledge and experience of individual forecasters. They are able to select relevant information from the real-time flux of observations and to interpret and diagnose existing conditions, all of which lead to their inference and decision making when issuing public advisories and warnings. Inevitably, capacity-building and ensuring the continuity of forecasting expertise constitute a great challenge for the NWS. This opens a wide range of opportunities for intelligence acquisition technology and the engineering of regional expert decision support systems to capitalize on the human intelligence at local WFOs and to make the most of artificial intelligence applications to process, classify, interpret, and synthesize inhomogeneous data.

Data Integration

A new data integration technology became available recently to WFOs as a component of AWIPS. The Flash Flood Monitoring and Prediction (FFMP; J. A. Smith et al., 2000; Filiaggi et al., 2002) application is a radar-based Advanced Hydrologic Prediction Services (AHPS; see Box 8.1) decision support tool that evolved from the Areal Mean Basin Estimated Rainfall

BOX 8.1

Advanced Hydrologic Prediction Services

The AHPS is a new program that was created as part of the modernization of the NWS hydrology program. AHPS is a web-based suite of forecast products that make use of modernized NWS technologies and improved meteorological and hydrological scientific understanding. The objective of the program is to provide improvements to river and flood forecasting capabilities and to improve communications between the NWS, cooperating agencies, and stakeholders to meet the diverse and evolving needs of all of these groups (NWS, 2001b). AHPS will achieve this objective by delivering several enhancements to the NWS Hydrologic Services Program (NWS, 2002b). Among these are

- better forecast accuracy—by incorporating advanced science into hydrologic modeling systems and coupling atmospheric and hydrologic models and forecast information on all time scales;
- more specific and timely information on fast-rising floods—by using tools to more rapidly identify small basins affected by heavy rainfall, identify excessive runoff locations, and predict the extent and timing of inundation;
- new types of forecast information—by incorporating new techniques for quantifying forecast certainty and conveying this information; and
- expanded outreach—by engaging partners and customers in all aspects of the hydrologic services improvement effort.

The National Research Council currently is reviewing the scope of the AHPS program, and it will produce a report by mid-2005. More information about the AHPS program can be found at *http://www.nws.noaa.gov/oh/ahps.*

(AMBER; Davis and Jendrowski, 1996) algorithm. This system assists local forecasters by monitoring precipitation accumulations at the hydrologic catchment scale to interpret a hydrologic threat within the context of an evolving meteorological situation to provide short-term, high-resolution flash flooding guidance. A prerequisite for utilization of FFMP is the topographic mapping of all the basins and subbasins covered by a WFO, including their hydrologic characteristics such as a basin's outline and average slope, soil characteristics and conditions, foliage coverage, and channel hydraulic measures. Time-varying parameters, such as soil moisture content or infiltration capacity, may be edited and updated by forecasters based on guidance received from the RFC.

The FFMP monitors initiation and movements of storms based on radar, satellite, and lightning data, and keeps track in real time of the rainfall

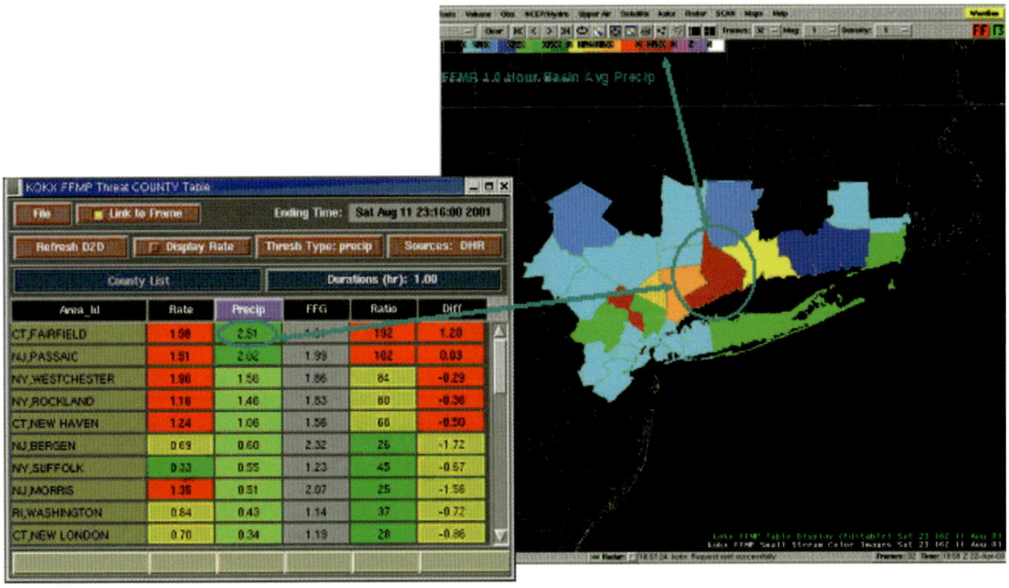

FIGURE 8.7 FFMP display showing precipitation and flooding information at the county level. SOURCE: National Weather Service.

accumulation and hydrologic conditions for each of the basins. The radar-estimated rainfall over a given catchment is operationally updated and the accumulation up to that point in time is compared to the respective flash flood guidance stored for that location. FFMP entails an interactive graphical display that enables the forecaster to monitor the responsible domain (Figure 8.7) and zoom in on any catchment by mouse click to retrieve current status and flash flood guidance information (Figure 8.8). The potential for flooding is color-coded to provide a quick assessment of where the situation may reach or exceed flash flood threat levels.

The high-resolution information and zoom-in capability and the automated bookkeeping of the FFMP system enable forecasters to stay abreast of potentially hazardous weather situations more easily, issue much more specific warnings, and do so with improved skills. Thus, FFMP bears the potential to greatly enhance flash flood monitoring, prediction, and warning. Although this technology is available, it has been implemented at only a few WFOs nationwide, because it requires a significant effort to digitize

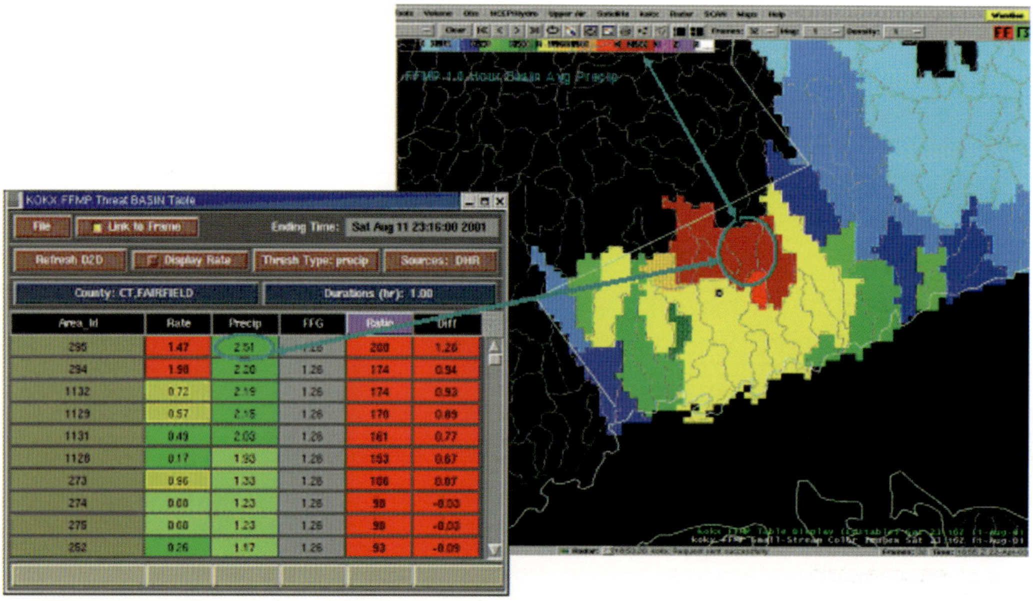

FIGURE 8.8 FFMP display showing precipitation and flooding information at the hydrologic basin level. SOURCE: National Weather Service.

the topographic and hydrologic information for each of the potentially thousands of basins and subbasins covered by a WFO. However, some agencies (e.g., county public works departments, water resource divisions) already have such topographic and hydrologic information for some of their watersheds; WFO collaboration with such agencies may somewhat lessen this burdensome task and facilitate the implementation of FFMP.

Future developments of FFMP may include incorporation of numerically predicted precipitation to increase the lead time in issuing flash flood warnings, which would be most valuable for regions of complex terrain.

Expert and Fusion Systems

The important concept behind data fusion and coupled hydrological-meteorological systems is to quickly combine relevant features in the observational and model datasets and produce a guidance product for the forecaster to view and edit if necessary (Roberts et al., 2003). The intent is to

Recommendation: In addition to the original NEXRAD siting considerations, future siting of radars in complex terrain should include detailed assessments of coverage in areas at risk for flash flooding.

The original NEXRAD siting procedures considered primarily meteorological processes, radar coverage, and ground clutter. Increased understanding of hydrologic processes of runoff production and streamflow response, combined with the application of radar to real-time hydrologic prediction for individual catchments (e.g., FFMP), enables incorporation of hydrologic aspects of the land surface into future siting processes. Readily available detailed digital information on topography and other relevant spatial information now make it possible to analyze a radar's coverage of the near-ground portion of the atmosphere using geographic information systems (GIS) technology. The potential for complete and partial radar beam blockage can be evaluated in the context of hydrologic basins for which coverage is sought. Basin size, average slope, orientation with respect to the movement of dominant weather patterns, characteristics of soil, land cover and land use, and channel hydraulic aspects determine the amount of rain that is likely to cause flash flooding and where it may occur. These characteristics, together with local hazard vulnerabilities, can help determine priorities of site selection.

save the forecaster significant time in assimilating all of the observational information by doing this automatically using a series of computer algorithms run every 6 to 10 minutes. This allows the forecaster more time to focus attention on the specific areas of concern for hazardous weather and to provide timely dissemination of warnings to the public. Most forecasters generally have high confidence in their ability to identify any threatening weather in the data. However, a human impacts study conducted by Anderson-Berry et al. (2004) on forecaster performance during the Sydney 2000 Olympics games showed that forecasters felt they could produce even better forecasts with access to more advanced technology and the automation of some products in order to give them more time to study the weather situation and local conditions. Furthermore, the Sydney forecasters felt that their warnings were superior to those they would have issued using routine storm tracking techniques. Thus, to advance the potential for providing improvements in flash flood forecasting, consideration should be given to employing the use of data assimilation and fusion systems in WFOs in the near future.

Elements of expert decision support systems already are present in the training modules available at WFOs and nationwide through AWIPS and IFPS user interface systems, which facilitate visualization and integration of multiple types of data (Meiggs et al., 1998; Ruth et al., 1998). The training modules consist of a limited number of cases at each WFO (e.g., the committee observed cases of heavy precipitation and flooding in Ventura County while at the Lox Angeles-Oxnard WFO), and the analysis software works on a sophisticated graphical display and analysis tool on a user prompt basis. The ability to retrieve historical case studies and to generate localized probabilistic products with 0- to 12-hour lead times, which will help stakeholders better understand uncertainty, would improve the quality and reliability of the support provided by IFPS to operational forecasters. This will continue to be important as FFMP and other expert decision support systems become widely available (Mass, 2003).

In addition to data assimilation methodologies, data fusion systems are being run operationally that combine different observational datasets together to produce short-term, 0- to 2-hour forecasts (or nowcasts) of precipitation rate and precipitation accumulation. One component of data fusion systems is the sophisticated feature detection and tracking algorithms that run on the radar (Dixon and Wiener, 1993; Wolfson et al. 1998; Germann and Zawadzki, 2002; Seed, 2003) and satellite (Evans and Shemo, 1996; Nesbitt et al., 2000; Roberts and Rutledge, 2003) data to monitor storm growth, evolution, and motion. NCAR runs an automated convective-storm nowcast system for the FAA, called the Auto-Nowcast System (ANC; Mueller et al., 2003), which produces 0- to 1-hour precipitation rate nowcasts every 6 minutes. In addition to nowcasting storm growth, evolution and decay, the ANC produces time- and location-specific nowcasts of storm initiation through the use of (1) Doppler radar to detect wind shifts or convergence boundaries at low altitudes; (2) satellite information to monitor the growth of cumulus clouds; (3) the VDRAS cloud model to obtain vertical motions in the boundary layer; and (4) national numerical model output to delineate regions of atmospheric instability (Saxen et al. 2004). Through the use of fuzzy logic (McNeill and Freiberger, 1993), the ANC is able to apply confidence values and add together the most important features detected in the observations to produce short-term convective storm nowcasts.

Additional systems that are operational include the National Convective Weather Forecast (NCWF; Megenhardt et al., 2000, 2004) system, which incorporates radar, lightning, and numerical model data, to produce

0- to 2-hour convective storm nowcasts and is currently in use at the Aviation Weather Center in Kansas City. Forecasters use this regularly updated (every 10 minutes), automated weather product as guidance in anticipating enroute aviation weather hazards. The United Kingdom Meteorology Office has run a system called Nimrod (Golding, 1998) operationally for several years. This system, which combines radar and satellite information with numerical model output, produces 0- to 6-hour precipitation rate and accumulation forecasts every 10 minutes for both meteorological and hydrological purposes; accurate warning for flash floods are a primary goal in the use of this tool.

Recommendation: To increase the accuracy and lead time of flash flood forecasts and warnings, the NWS should continue to implement new technologies and techniques including (a) the Flash Flood Monitoring and Prediction program at all Weather Forecast Offices, (b) polarimetric modifications to NEXRAD, (c) data assimilation systems that integrate radar and other operational datasets into coupled hydrometeorological and hydrological models, and (d) data fusion systems.

Extensive opportunities exist for forecasters to take advantage of the rapid advancement of technology to improve forecasts, watches, and warnings. The FFMP system, which requires adaptation to the specific watersheds served by each WFO, would facilitate more specific flash flood warnings. In addition, as part of its new Advanced Hydrologic Prediction Services, the NWS is encouraged to continue its effort to develop and evaluate hydrologic and coupled meteorological-hydrologic models to advance technologies useful for improved flash flood guidance and warnings. The polarimetric modification would improve the data quality and quantitative precipitation measurement capabilities of NEXRAD. Real-time data assimilation systems that incorporate observations into high-resolution mesoscale numerical models provide rapidly updated wind and precipitation forecasts. Data fusion systems, such as the Auto-Nowcast system, incorporate all available observation datasets together with numerical model output to produce very short range (0- to 2-hour), site-specific forecasts. These advances can produce improved forecasts, including ensemble and probabilistic forecasts, of precipitation rate and accumulation that are essential for flash flood forecasting. To enhance the usefulness of the forecast, quantification of uncertainty (e.g., probability forecasts) should be an integral component of these products (NRC, 2003).

9

Concluding Thoughts

The committee has reviewed a great deal of material related to flash flood forecasting and warning and NEXRAD coverage—mostly dealing with the specific case of the Los Angeles-Oxnard (LOX) Weather Forecast Office (WFO) and the Sulphur Mountain NEXRAD. The overall strategy of using NEXRAD to monitor the development, evolution, and movement of storms and to estimate the resulting precipitation over the radar coverage area seems well suited to the flash flood forecasting and warning mission of the NWS. The principal shortcoming of the system for radars sited in complex terrain, such as that around Los Angeles, is the essentially unavoidable gaps in radar coverage. These gaps are due primarily to the effects of Earth's curvature or blocking of the radar beam by intervening terrain, and the consequent inability of the beam to reach down into some low-lying areas.

The current restriction of the NEXRAD scans to a minimum elevation angle of 0.5° exacerbates the problem. From the Sulphur Mountain radar altitude of 831 m (2726 ft) the axis of the radar beam at 0.5° elevation angle passes above 1.83-km (6000-ft) altitude beyond a distance of about 75 km from the radar site. The lower edge of the beam, as defined by the half-power or 3-dB point in the antenna pattern, passes above that altitude beyond about 125 km. Thus, precipitation below 6000 ft can be detected only within about 100–125 km from the Sulphur Mountain radar. This problem of detecting low-level precipitation is even worse for other NEXRADs, mostly in the western United States, which are sited at even higher altitudes than the Sulphur Mountain radar.

Despite the shortcomings posed by its elevated site, the Sulphur Mountain radar, in conjunction with its neighboring NEXRADs, consistently has detected heavy precipitation threatening the county warning areas served by the LOX WFO. The radar availability has exceeded, with rare exception, the National Weather Service (NWS) goal of greater than 96 percent availability over the last several years. The LOX WFO record of flash flood warnings has been better in most respects than the national median performance, and the national database mistakenly includes some "missed events" that either were not flash flood cases or were actually covered by timely warnings. An apparent shortcoming of the LOX office in its average warning lead time compared to the national average is misleading, because warning times tend to be shorter in steep terrain such as that around the Los Angeles area. That is reflected in the NWS Western Region 2004 goal for flash flood warning lead time, which the LOX WFO has routinely exceeded. Based on all of these analyses, ***the committee finds that the Sulphur Mountain radar is appropriately sited to detect approaching storms while avoiding problems with anomalous propagation of the radar signals. The radar is amply functional and has provided crucial support to the Los Angeles-Oxnard forecasters in their mission to monitor, predict, and warn of precipitating events and flash floods.***

Nonetheless, it is clear that better low-level coverage can be achieved from the Sulphur Mountain radar as well as other mountaintop radar sites. The straightforward use of lower minimum elevation angles in the scans would provide coverage down to lower altitudes at greater ranges in any direction not blocked by terrain. In the Sulphur Mountain case, this would extend the low-level coverage farther out over the ocean to the southwest, from where many winter storms approach, and over the main part of Los Angeles to the southeast. At 0.0° elevation, for example, the beam axis would be below 1.83 km (6000 ft) out to 125 km, while the lower edge of the beam would be below that level to well beyond 150 km. This would clearly enhance the ability to sense low-level precipitation to a greater range and should help improve the flash flood warning capability of the LOX office (e.g., by permitting warnings with greater lead times). Similar results can be anticipated for other mountaintop NEXRAD sites, and even slightly negative elevation angles would be useful in many cases.

Finally, further improvements in flash flood warning capabilities are on the way, in both the radar and the flash flood forecasting arenas. NEXRAD precipitation products are continually being improved, and the forthcoming polarimetric modification to NEXRAD will improve rainfall estimates, especially in directions where the beam is partially blocked. Moreover, other technologies and instrumentation (e.g., phased-array radars) are being researched and tested; as these become feasible for use in the future, they should be brought to bear on the flash flood forecasting problem as appropriate. In addition, the Flash Flood Monitoring and Prediction (FFMP) program, yet to be adapted for the area served by the LOX office, will help to improve the skill and areal specificity in flash flood forecasting. These and other forthcoming developments promise to improve the already fine record of the LOX office (as well as others) in the flash flood warning efforts.

References

Alberoni, P. P., T. Andersson, P. Mezzasalma, D. B. Michelson, and S. Nanni. 2001. Use of the vertical reflectivity profile for identification of anomalous propagation. *Meteorological Applications*, **8**, 257-266.

AMS (American Meteorological Society). 2000. *Glossary of Meteorology*. T. S. Glickman, ed., 2nd Edition. Boston: American Meteorological Society. 855 pp.

Amitai, E. 1999. Relationships between radar properties at high elevations and surface rain rate: Potential use for spaceborne rainfall measurements. *Journal of Applied Meteorology*, **38**, 321-333.

Anagnostou, E. N., W. F. Krajewski, D. J. Seo, and E. R. Johnson. 1998. Mean field radar rainfall bias studies for WSR-88D. *ASCE Journal of Engineering Hydrology*, **3**(3), 149-159.

Anagnostou, E. N., W. F. Krajewski, and J. A. Smith. 1999. Uncertainty quantification of mean-areal radar-rainfall estimates. *Journal of Atmospheric and Oceanic Technology*, **16**, 206-215.

Anagnostou, E. N., M. N. Anagnostou, W. F. Krajewski, A. Kruger, and B. J. Miriovsky. 2004. High-resolution rainfall estimation from X-band polarimetric radar measurements. *Journal of Hydrometeorology*, **5**, 110-128.

Anderson-Berry, L., T. Keenan, J. Bally, R. Pielke, Jr., R. Leigh, and D. King. 2004. The societal, social and economic impacts of the World Weather Research Program Sydney 2000 Forecast Demonstration Project (WWRP S2000 FDP). *Weather and Forecasting*, **19**, 168-178.

Andrews, E. D., R. C. Antweiler, P. J. Neiman, and F. M. Ralph. 2004. Influence of ENSO on flood frequency along the California coast. *Journal of Climate*, **17**, 337-348.

Applequist, S., G. E. Gahrs, R. L. Pfeffer, and X. Niu. 2002. Comparison of methodologies for probabilistic quantitative precipitation forecasting. *Weather and Forecasting*, **17**(4), 783-799.

Austin, P. M. 1987. Relation between measured radar reflectivity and surface rainfall. *Monthly Weather Review*, **115**, 1053-1071.

Babin, S. M. 1996. Surface duct height distributions for Wallops Island, Virginia, 1985-1994. *Journal of Applied Meteorology*, **35**, 86-93.

Babin, S. M., and J. R. Rowland. 1992. Observation of a strong surface radar duct using helicopter acquired fine-scale radio refractivity measurements. *Geophysical Research Letters*, **19**, 917-920.

Bae, D. H., K. P. Georgakakos, and S. K. Nanda. 1995. Operational forecasting with real-time databases. *ASCE Journal of the Hydraulics Division*, **121**, 49-60.

Baeck, M. L., and J. A. Smith. 1998. Rainfall estimation for the WSR-88D for heavy rainfall events. *Weather and Forecasting*, **13**, 416-436.

Baer, V. E. 1991: The transition from the present radar dissemination system to the NEXRAD Information Dissemination Service (NIDS). *Bulletin of the American Meteorological Society*, **72**, 29-33.

Balakrishnan, N., and D. S. Zrnic. 1990. Estimation of rain and hail rates in mixed-phase precipitation. *Journal of the Atmospheric Sciences*, **47**, 565-583.

Barros, A. P., and R. J. Kuligowski. 1998. Orographic effects during a severe wintertime rainstorm in the Appalachian Mountains. *Monthly Weather Review*, **126**, 2648-2672.

Bean, B., and E. Dutton. 1966. *Radio Meteorology*. National Bureau of Standards, Monograph 92. Washington, D.C.: U.S. Government Printing Office. 425 pp.

Bech, J., B. Codina, J. Lorente, and D. Bebbington. 2003. The sensitivity of single polarization weather radar beam blockage correction to variability in the vertical refractivity gradient. *Journal of Atmospheric and Oceanic Technology*, **20**(6), 845-855.

Bech, J., A. Sairouni, B. Codina, J. Lorente, and D. Bebbington. 2000. Weather radar anaprop conditions at a Mediterranean coastal site. *Physics and Chemistry of the Earth. Part B—Hydrology, Oceans and Atmosphere*, **25**, 829-832.

Bell, R. S. 1978. The forecasting of orographically enhanced rainfall accumulations using 10-level model data. *Meteorological Magazine*, **107**, 113-124.

Beven, K. J. 2001. *Rainfall-Runoff Modelling: The Primer*. New York: John Wiley & Sons, 360 pp.

Bluestein, H. B., B. A. Albrecht, R. M. Hardesty, W. D. Rust, D. Parsons, R. Wakimoto, and R. M. Rauber. 2001. Ground-based mobile instrument workshop summary, 23-24 February 2000, Boulder, Colorado. *Bulletin of the American Meteorological Society*, **82**, 681-694.

Bond, N. A., C. F. Mass, B. F. Smull, R. A. Houze, M.-J. Yang, B. A. Colle, S. A. Braun, M. A. Shapiro, B. R. Colman, P. J. Neiman, J. E. Overland, W. D. Neff, and J. D. Doyle. 1997. The coastal observation and simulation with topography (COAST) experiment. *Bulletin of the American Meteorological Society*, **78**, 1941-1955.

Bougeault, P., P. Binder, A. Buzzi, R. Dirks, R. A. Houze, J. Kuettner, R. B. Smith, R. Steinacker, and H. Volkert. 2001. The MAP special observing period. *Bulletin of the American Meteorological Society*, **82**, 433-462.

Bradley, A. A., and K. W. Potter. 1992. Flood frequency analysis of simulated flows. *Water Resources Research*, **28**, 2375-2385.

Brandes, E. A., A. V. Ryzhkov, and D. S. Zrnic. 2001. An evaluation of radar rainfall estimates from specific differential phase. *Journal of Atmospheric and Oceanic Technology*, **18**, 363-375.

Brandes, E. A., G. Zhang, and J. Vivekanandan. 2002. Experiments in rainfall estimation with a polarimetric radar in a subtropical environment. *Journal of Applied Meteorology*, **41**, 674-685.

Braun, S. A., R. A. Houze, Jr., and B. F. Smull. 1997. Airborne dual-Doppler observations of an intense frontal system approaching the Pacific Northwest coast. *Monthly Weather Review*, **125**, 3131-3156.

Bringi, V. N., and V. Chandrasekar. 2001. *Polarimetric Doppler Weather Radar—Principles and Applications*. Cambridge, U.K.: Cambridge University Press.

Bringi, V. N., V. Chandrasekar, J. Hubbert, E. Gorgucci, W. L. Randeu, and M. Schoenhuber. 2003. Raindrop size distribution in different climatic regimes from disdrometer and dual-polarized radar analysis. *Journal of the Atmospheric Sciences*, **60**, 354-365.

Brooks, I. M., A. K. Goroch, and D. P. Rogers. 1999. Observations of strong surface radar ducts over the Persian Gulf. *Journal of Applied Meteorology*, **38**, 1293-1310.

Brown, R. A., V. T. Wood, and T. Barker. 2002. Improved detection using negative elevation angles for mountaintop WSR-88Ds: Simulation of KMSX near Missoula, Montana. *Weather and Forecasting*, **17**(2), 223-237.

Browning, K. A., C. W. Pardoe, and F. F. Hill. 1975. The nature of orographic rain at wintertime cold fronts. *Quarterly Journal of the Royal Meteorological Society*, **101**, 333-352.

Browning, K. A., F. F. Hill, and C. W. Pardoe. 1974. Structure and mechanism of precipitation and the effect of orography in a wintertime warm sector. *Quarterly Journal of the Royal Meteorological Society*, **100**, 309-330.

Bruintjes, R. T., T. L. Clark, and W. D. Hall. 1994. Interactions between topographic airflow and cloud/precipitation development during the passage of a winter storm in Arizona. *Journal of the Atmospheric Sciences*, **51**, 48-67.

BAMS (Bulletin of the American Meteorological Society). 2000. Policy statement: Prediction and mitigation of flash floods. **81**(6), 1338-1340.

Burk, S. D., and W. T. Thompson. 1997. Mesoscale modeling of summertime refractive conditions in the Southern California bight. *Journal of Applied Meteorology*, **36**, 22-31.

Burns, P. L., J. Petriceks, R. V. Schledewitz, G. K. Brown, L. Hawke-Gerrans, and J. A. Chamberlain. 1987: SRI NEXRAD In-Depth Site Survey Report for the Los Angeles, California, Area (Sulphur Mountain Site). DOC, NEXRAD JSPO, Silver Spring, MD.

Buzzi, A., N. Tartaglione, and P. Malguzzi. 1998. Numerical simulations of the 1994 Piedmont flood: Role of orography and moist processes. *Monthly Weather Review*, **126**, 2369-2383.

CADOF (California Department of Finance). 2002. California statistical abstract. Available online at *http://www.dof.ca.gov/html/fs_data/stat-abs/CA_StatAbs02w.pdf* or from the California Department of Finance, 915 L Street, Sacramento, Calif.

Caracena, F., R. A. Maddox, L. R. Hoxit, and C. F. Chappell. 1979. Mesoanalysis of the Big Thompson storm. *Monthly Weather Review*, **107**, 1-17.

Cayan, D. R., K. T. Redmond, and L. G. Riddle. 1999. ENSO and hydrologic extremes in the western United States. *Journal of Climate*, **12**, 2881-2893.

Chandrasekar, V., V. N.Balakrishnan, and D. S. Zrnic. 1990. Error structure of multiparameter radar and surface measurements of rainfall. Part III: Specific differential phase. *Journal of Atmospheric and Oceanic Technology*, **7**, 621-629.

Chandrasekar, V., W. A. Cooper, and V. N. Bringi. 1988. Axis ratios and oscillations of raindrops. *Journal of the Atmospheric Sciences*, **45**, 1323-1333.

Chandrasekar, V. and V. N. Bringi. 2004. Dual polarization radar estimates of rainfall: Recent advances. Proceedings of the Sixth International Symposium on Hydrological Applications of Weather Radar, Melbourne, Australia, pp. 1-6.

Chappell, C. F. 1986. Quasi-stationary convective events. Pp. 289-310 in *Mesoscale Meteorology and Forecasting*, Peter S. Ray, ed. Boston: American Meteorological Society.

Charalampidis, D., T. Kasparis, and W. L. Jones. 2002. Removal of nonprecipitating echoes in weather radar using multifractals and intensity. *IEEE Transactions on Geoscience and Remote Sensing*, **40**, 1121-1131.

Chen, F., T. Warner, and K. Manning. 2001. Sensitivity of orographic moist convection to landscape variability: A study of the Buffalo Creek, Colorado, flash flood cast of 1996. *Journal of the Atmospheric Sciences*, **58**(12), 3204-3223.

Clapp, R. B., and G. M. Hornberger. 1978. Empirical equations for some soil hydraulic properties. *Water Resources Research*, **14**, 601-604.

Colle, B. A. 2004. Sensitivity of orographic precipitation to changing ambient conditions and terrain geometries: An idealized modeling perspective. *Journal of the Atmospheric Sciences*, **61**, 588-606.

Colle, B. A., and C. F. Mass. 1996. An observational and modeling study of the interaction of low-level southwesterly flow with the Olympic Mountains during COAST IOP 4. *Monthly Weather Review*, **124**, 2152-2175.

Colle, B. A., and C. F. Mass, 2000: The 5-9 February 1996 flooding event over the Pacific Northwest: Sensitivity studies and evaluation of the MM5 precipitation forecasts. *Monthly Weather Review*, **128**, 593-617.

Collier, C. G. 1975. A representation of the effects of topography on surface rainfall within moving baroclinic disturbances. *Quarterly Journal of the Royal Meteorological Society*, **101**, 407-422.

Cook, J. 1991. A sensitivity study of weather data inaccuracies on evaporation duct height algorithms. *Radio Science*, **26**, 731-746.

Corrigan, P., D. D. Fenn, D. R. Kluck, and J. J. Vogel. 1999. Probable maximum precipitation for California. Hydrometeorological Report No. 59. Silver Spring, Md.: National Weather Service.

Cortinas, J. V., and D. J. Stensrud. 1995. The importance of understanding mesoscale model parameterization schemes for weather forecasting. *Weather and Forecasting,* **10**, 716-740.

Cotton, W. R., and R. A. Anthes. 1989. *Storm and Cloud Dynamics*. Burlington, Mass.: Academic Press, 158 pp.

Cress, A., E. Heise, and G. Rivin. 2002. High-resolution QPF and precipitation analyses for use in hydrological models. Abstracts, Vol. 2, International Conference on Quantitative Precipitation Forecasting, Reading, UK. 120 pp.

Crum, T. D., and R. L. Alberty. 1993. The WSR-88D and the WSR-88D operational support facility. *Bulletin of the American Meteorological Society,* **74**, 1669-1687.

Crum, T. D., R. L. Alberty, and D. W. Burgess. 1993. Recording, archiving, and using WSR-88D data. *Bulletin of the American Meteorological Society,* **74**(4), 645-653

Crum, T. D., R. E. Saffle, and J. W. Wilson. 1998. An update on the NEXRAD program and future WSR-88D support to operations. *Weather and Forecasting,* **13**, 253-262.

da Silveira, R. B., and A. R. Holt. 2001. An automatic identification of clutter and anomalous propagation in polarization-diversity weather radar data using neural networks. *IEEE Transactions on Geoscience and Remote Sensing,* **39**, 1777-1788.

Davis, R. D., and P. Jendrowski. 1996. The operational Areal Mean Basin Estimated Rainfall (AMBER) module. *Preprints, 16th Conference on Weather Analysis and Forecasting,* American Meteorological Society, pp. 332-335.

Davis, R. S. 2001. Flash flood forecast and detection methods. *Severe Convective Storms—AMS Meteorological Monograph Series,* **28**(50), 481-525.

Dixon, M., and G. Wiener. 1993. TITAN: Thunderstorm identification, tracking, analysis and nowcasting—a radar-based methodology. *Journal of Atmospheric and Oceanic Technology,* **10**, 785-797.

Dolcine, L., H. Andrieu, and M. N. French. 1998. Evaluation of a conceptual rainfall forecasting model from observed and simulated rain events. *Hydrology and Earth System Sciences,* **2**, 173-182.

Done, J., C. Davis, and M. Weisman. 2004. The Next Generation of NWP: Explicit forecasts of convection using the Weather Research and Forecasting (WRF) model. *Atmospheric Science Letters,* in press.

Doswell, C. A. III, H. E. Brooks, and R. A. Maddox. 1996. Flash flood forecasting: an ingredients-based methodology. *Weather and Forecasting,* **11**, 560-581.

Doviak, R. J. and D. S. Zrnic. 1985. Siting of Doppler weather radars to shield ground targets. *IEEE Transactions, Antennas, and Propagation,* **AP-33**(7), 685-689.

Doviak, R. J., V. Bringi, A. Ryzhkov, A. Zahrai, and D. Zrnic. 2000. Considerations for polarimetric upgrades to operational WSR-88D radars. *Journal of Atmospheric and Oceanic Technology,* **17**(3), 257-278.

Doyle, J. D. 1997. The influence of mesoscale orography on a coastal jet and rainband. *Monthly Weather Review,* **125**, 1465-1488.

Droegemeier, K. K., K. Kelleher, T. Crum, J. J. Levit, S. A. Del Greco, L. Miller, C. Sinclair, M. Benner, D. W. Fulker, and H. Edmon. 2002. Project CRAFT: A test bed for demonstrating the real time acquisition and archival of WSR-88D level II data. *Preprints, 18th International Conference on Interactive Information Processing Systems (IIPS) for Meteorology, Oceanography, and Hydrology,* American Meteorological Society.

Du, J., S. L. Mullen, and F. Sanders. 1997. Short-range ensemble forecasting of quantitative precipitation. *Monthly Weather Review,* **125**, 2427-2459.

Ebert, E. E. 2001. Ability of a poor man's ensemble to predict the probability and distribution of precipitation. *Monthly Weather Review*, **129**, 2461-2480.

Eichinger, W. E., D. I. Cooper, M. Parlange, and G. Katul. 1993. The application of a scanning, water Raman-lidar as a probe of the atmsopheric boundary layer. *IEEE Transactions on Geoscience and Remote Sensing*, **31**, 70-79.

Eichinger, W. E., D. I. Cooper, P. R. Forman, J. Griegos, M. A. Osborn, D. Richter, L. L. Tellier, and R. Thornton. 1999. The development of a scanning Raman water vapor lidar for boundary layer and tropospheric observations. *Journal of Atmospheric and Oceanic Technology*, **16**, 1753-1766.

Evans, J. L., and R. E. Shemo. 1996. A procedure for automated satellite-based identification and climatology development of various classes of organized convection. *Journal of Applied Meteorology*, **35**, 638-652.

Evans, J. E., K Carusone, M. Wolfson, B. Crowe, D. Meyer, and D. Klingle-Wilson. 2002. The Corridor Integrated Weather System (CIWS). *Preprints, 10th Conference on Aviation, Range and Aerospace*, American Meteorological Society, pp. 210-215.

Fabry, F. 2004. Meteorological value of ground target measurements by radar. *Journal of Atmospheric and Oceanic Technology*, **21**, 560-573.

Fabry, F., C. Frush, I. Zawadzki, and A. Kilambi. 1997. On the extraction of near-surface index of refraction using radar phase measurements from ground targets. *Journal of Atmospheric and Oceanic Technology*, **14**, 978-987.

Filiaggi, M. T., S. B. Smith, M. Churma, L. Xin, and M. Glaudemans. 2002. Flash Flood Monitoring and Prediction version 2.0: Continued AWIPS modernization. *Preprints, 18th International Conference on Interactive Information Processing Systems (IIPS) for Meteorology, Oceanography, and Hydrology*, American Meteorological Society.

Fisk, C. 1994: *A Climatology of the San Nicolas Island Temperature Inversion*. Geophysics Division Technical Note No. 181. Point Mugu, Calif.: Naval Air Warfare Center Weapons Division, 33 pp.

Fox, A. 2003. Using NexRad radar in Southern California to monitor rain: Some insights from an operational forecasting perspective. Port Hueneme, Calif.: Fox Weather, 16 pp.

French, M. N., and W. F. Krajewski. 1994. A model for real-time quantitative rainfall forecasting using remote sensing. Part I. Formulation. *Water Resources Research*, **30**, 1075-1083.

Friday, E. W. 1994. The modernization and associated restructuring of the National Weather Service: An overview. *Bulletin of the American Meteorological Society*, **75**, 43-52.

Fritsch, J. M., and R. E. Carbone. 2004. Improving quantitative precipitation forecasts in the warm season: A USWRP research and development strategy. *Bulletin of the American Meteorological Society*, **85**(7), 955-965.

Fujita, T. T. 1981. Tornadoes and downbursts in the context of generalized planetary scales. *Journal of the Atmospheric Sciences*, **38**, 1511-1534.

Fulton, R. A., J. P. Breidenbach, D. J. Seo, D. A. Miller, and T. O'Bannon. 1998. The WSR-88D rainfall algorithm. *Weather and Forecasting*, **13**(2), 377-395.

Ganguly, A. R., and R. L. Bras. 2003. Distributed quantitative precipitation forecasting using information from radar and numerical weather prediction models. *Journal of Hydrometeorology*, **4**, 1168-1180.

GAO (General Accounting Office). 1998. *National Weather Service: Sulphur Mountain Radar Performance*. GAO/AIMD-99-7. Washington, D.C.: U.S. General Accounting Office, 34 pp.

Georgakakos, K. P., and M. D. Hudlow. 1984. Quantitative precipitation forecasting techniques for use in hydrologic forecasting. *Bulletin of the American Meteorological Society*, **65,** 1186-1200.

Georgakakos, K. P. 1986a. On the design of national, real-time warning systems with capability for site-specific, flash-flood forecasts. *Bulletin of the American Meteorological Society*, **67**, 1233-1239.

Georgakakos, K. P. 1986b. A generalized stochastic hydrometeorological model for flood and flash-flood forecasting. Part I. Formulation. *Water Resources Research*, **22**, 2083-2095.

Georgakakos, K. P. 1986c. A generalized stochastic hydrometeorological model for flood and flash-flood forecasting. Part II. Case studies. *Water Resources Research*, **22**, 2096-2106.

Georgakakos, K. P. 1987. Real-time flash flood prediction. *Journal of Geophysical Research*, **92**, 9615-9629.

Georgakakos, K. P. 2002. U.S. corporate technology transfer in hydrometeorology. *Journal of Hydroinformatics*, **4**, 3-13.

Georgakakos, K. P., and E. Foufoula-Georgiou. 1991. Real-time coupling of hydrologic and meteorological models for flood forecasting. Pp. 169-184 in *Recent Advances in the Modeling of Hydrologic Systems*. D. S. Bowles and P. E. O'Connell, eds. Boston, Mass.: Kluwer Academic Publishers.

Georgakakos, K. P., and M. D. Hudlow. 1984. Quantitative precipitation forecast techniques for use in hydrologic forecasting. *Bulletin of the American Meteorological Society*, **65**, 1186-1200.

Germann, U., and J. Joss. 2002. Mesobeta profiles to extrapolate radar precipitation measurements above the Alps to the ground level. *Journal of Applied Meteorology*, **41**, 542-557.

Germann, U., and I. Zawadzki. 2002. Scale-dependence on the predictability of precipitation from continental radar images. Part I: Description of the methodology. *Monthly Weather Review*, **130**, 2859-2873.

Germann, U., and I. Zawadzki. 2004: Scale dependence of the predictability of precipitation from continental radar images. Part II: Probability forecasts. *Journal of Applied Meteorology*, **43**, 74-89.

Glahn, H. R., and D. P. Ruth. 2003. The new Digital Forecast Database of the National Weather Service. *Bulletin of the American Meteorological Society*, **84**, 195-201.

Golding, B. W. 1998. Nimrod: A system for generating automated very short range forecasts. *Meteorological Applications*, **5**, 1-16.

Gossard, E. E. 1977. Refractive index variance and its height distribution in different air masses. *Radio Sciences*, **12**, 89-105.

Gossard, E. E., R. G. Strauch, B. B. Stankov, and D. E. Wolfe. 1995. Measurement of property gradients and turbulence aloft with ground-based Doppler radars. NOAA Technical Memorandum ERL 453-ETL 67. Boulder, Colo.: Environmental Technology Laboratory, 31 pp.

Graf, W. L. 1977. Network characteristics in suburbanizing streams. *Water Resources Research*, **13**, 459-463.

Grimit, E. P., and C. F. Mass. 2002. Initial results of a mesoscale short-range ensemble forecasting system over the Pacific Northwest. *Weather and Forecasting*, **17**, 192-205.

Gruntfest, E., and C. J. Huber. 1991. Toward a comprehensive national assessment of flash flooding in the United States. *Episodes*, **14**, 26-35.

Gumbrecht, B. 1999. The Los Angeles River: Its Life, Death and Possible Rebirth. Baltimore: Johns Hopkins University Press.

Habib, E., G. J. Ciach, and W. F. Krajewski. 2004. A method for filtering out raingauge representativeness errors from the verification distributions of radar and raingauge rainfall. *Advances in Water Resources*, in press.

Hall, T., H. Brooks, and C. D. Doswell III. 1999. Precipitation forecasting using a neural network. *Weather and Forecasting*, **14**(3), 338-345.

Harris, D., E. Foufoula-Georgiou, K. K. Droegemeier, and J. J. Levit. 2001. Multiscale statistical properties of a high-resolution precipitation forecast. *Journal of Hydrometeorology*, **2**, 406-418.

Haynes, A. 2001. Synoptic pattern typing for historical heavy precipitation events in southern California. Western Regional Technical Attachment No. 01-15, National Weather Service. Available online at *http://www.wrh.noaa.gov/wrhq/01TAs/0115/index.html.*

Heiss, W. H., D. L. McGrew, and D. Sirmans. 1990. NEXRAD: Next Generation Weather Radar (WSR-88D). *Microwave Journal*, **33**, 79-98.

Hirschboeck, K. K. 1987. Catastrophic flooding and atmospheric circulation anomalies. Pp. 23-56 in *Catastrophic Flooding*, L. Mayer and D. Nash, Eds. Boston: Allen and Unwin.

Hjelmfelt, M. R. 1999. Flash floods associated with convective storms. Pp. 80-102 in *Storms*, Roger Pielke Sr. and Roger Pielke Jr., eds. New York: Routledge Press.

Hobbs, P. V., and O. G. Persson. 1982. The mesoscale and microscale structure and organization of clouds and precipitation in midlatitude cyclones. Part V: The substructure of narrow cold-frontal rainbands. *Journal of the Atmospheric Sciences*, **39**, 280-295.

Hobbs, P. V., R. A. Houze, and T.J. Matejka. 1975. The dynamic and microphysical structure of an occluded front and its modification by orography. *Journal of the Atmospheric Sciences*, **32**, 1542-1562.

Hollis, G. E. 1988. Rain, roads, roofs, and runoff: Hydrology in cities. *Geography*, **73**, 9-18.

Horritt, M. S., and P. D. Bates. 2001. Predicting floodplain inundation: Raster-based modeling versus the finite-element approach. *Hydrological Processes*, **15**, 825-842.

Houze, R. A., Jr., J. D. Locatelli, and P. V. Hobbs. 1976. Dynamics and cloud microphysics of the rainbands in an occluded frontal system. *Journal of the Atmospheric Sciences*, **33**, 1921-1936.

Huber, C. J. 1992. Comparative assessment of flood warning systems in the United States, 1987 to 1991. Unpublished master's thesis in Geography, University of Colorado, Boulder.

Hughes, D. A. 1980. Floodplain inundation—Processes and relationships with channel discharge. *Earth Surface Processes and Landforms*, **5**, 297-304.

Illingworth, A. J., and T. M. Blackman. 2002. The need to represent raindrop size spectra as normalized gamma distributions for the interpretation of polarization radar observations. *Journal of Applied Meteorology*, **41**, 286-297.

Jameson, A. R. 1989. The interpretation and meteorological application of radar backscatter amplitude ratios at linear polarizations. *Journal of Atmospheric and Oceanic Technology*, **6**, 908-919.

Jameson, A. R. 1991. The effect of drop-size distribution variability on radiometric estimates of rainfall rates for frequencies from 3 to 10 GHz. *Journal of Applied Meteorology*, **30**, 1025-1033.

Jameson, A. R., and A. B. Kostinski. 1999. Non-Rayleigh signal statistics in clustered statistically homogeneous rain. *Journal of Atmospheric and Oceanic Technology*, **16**, 575-583.

Johnson, L. E. 2000. Assessment of flash flood warning procedures. *Journal of Geophysical Research*, **105D**, 2299-2313.

Johnson, L. E., W. F. Roberts, B. E. Skahill, and S. O'Donnell. 2001. Predictive uncertainty of watershed rainfall-runoff models using weather radar. *Preprints, 30th International Conference on Radar Meteorology*, American Meteorological Society, pp. 11-13.

Jordan, P. W., A. W. Seed, and P. E. Weinmann. 2003. A stochastic model of radar measurement errors in rainfall accumulations at catchment scale. *Journal of Hydrometeorology*, **4**, 841-855.

Joss, J., and R. Lee. 1995. The application of radar-gauge comparisons to operational precipitation profile corrections. *Journal of Applied Meteorology*, **34**, 2612-2630.

Joss, J., and A. Pittini. 1991. Real-time estimation of the vertical profile of radar reflectivity to improve the measurement of precipitation in an Alpine region. *Meteorology and Atmospheric Physics*, **47**, 61-72.

Jothityangkoon, C., and M. Sivapalan. 2003. Towards estimation of extreme floods: Examination of the roles of runoff process changes and floodplain flows. *Journal of Hydrology*, **281**, 206-229.

Journal of Hydrology. 2004. **298**(1-4).

Junker, N. W. 1992. *Heavy Rain Forecasting Manual.* Kansas City, Mo.: National Weather Service Training Center, 91 pp.

Kelsch, M. 2001. Hydrometeorological characteristics of flash floods. Pp. 181-193 in *Coping with Flash Floods*, E. Gruntfest and J. Handmer, eds. NATO Science Series 2. Environmental Security—Vol. 77. Boston: Kluwer Academic Publishers.

Kidder, S. Q., and T. H. Vonder Haar. 1995. *Satellite Meteorology: An Introduction*. Burlington, Mass.: Academic Press, 466 pp.

Kim, G., and A. P. Barros. 2001. Quantitative flood forecasting using multisensor data and neural networks. *Journal of Hydrology*, **246**, 45-62.

Klazura, G. E., J. M. Thomale, D. S. Kelly, and P. Jendrowski. 1999. A comparison of NEXRAD WSR-88D radar estimates of rain accumulation with gauge measurements for high- and low-reflectivity horizontal gradient precipitation events. *Journal of Atmospheric and Oceanic Technology*, **16**, 1842-1850.

Kostinski, A. B. 1994. Fluctuations of differential phase and radar measurements of precipitation. *Journal of Applied Meteorology*, **33**, 1176-1181.

Krajewski, W. F., and J. A. Smith. 2002. Radar hydrology: rainfall estimation. *Advances in Water Resources,* **25** (8-12), 1387-1394

Krajewski, W. F., and B. Vignal. 2001. Evaluation of anomalous propagation echo detection in WSR-88D data: A large sample case study. *Journal of Atmospheric and Oceanic Technology*, **18**, 807-814.

Krishnamurti, T. N., C. M. Kishtawal, Z. Zhang, T. LaRow, D. Bachiochi, E. Williford, S. Gadgil, and S. Surendran. 2000. Multimodel ensemble forecasts for weather and seasonal climate. *Journal of Climate*, **13**, 4196-4216.

Krzysztofowicz, R., W. J. Drzal, T. Rossi Drake, J. C. Weyman, and L. A. Giordano. 1993. Probabilistic Quantitative Precipitation Forecasts for River Basins. *Weather and Forecasting,* **8**(4), 424-439.

Kucera, P. A., W. F. Krajewski, and C. B. Young. 2004. Radar beam occultation studies using GIS and DEM technology: An example study of Guam. *Journal of Oceanic and Atmospheric Technology*, **21**(7), 995-1006.

Kuligowski, R. J., and A. P. Barros. 1998. Experiments in Short-term Precipitation Forecasting Using Artificial Neural Networks. *Monthly Weather Review*, **126**, 470-482.

LADPW (Los Angeles County Department of Public Works). 2004. Precipitation. Available online at *http://ladpw.org/wrd/precip/index.cfm.* Accessed May 7, 2004.

Landel, G., J. A. Smith, M. L. Baeck, M. Steiner, and F. L. Ogden. 1999. Radar studies of heavy convective rainfall in mountainous terrain. *Journal of Geophysical Research*, **104D**, 31451-31465.

Lee, T. H., and K. P. Georgakakos. 1996. Operational rainfall prediction on meso-gamma scales for hydrologic applications. *Water Resources Research*, **32**, 987-1003.

Leone, D. A., R. M. Endlich, J. Petriceks, R. Collis, and J. R. Porter. 1989. Meteorological considerations used in planning the NEXRAD Network. *Bulletin of the American Meteorological Society*, **70**, 4-13.

Leone, D., and M. H. Johnson. 1986. SRI NEXRAD Preliminary Site Survey for the Los Angeles, California Area (Sulphur Mountain Site). DOC, NEXRAD JSPO, Silver Spring, MD.

Leopold, L. B. 1968. *Hydrology for Urban Land Planning—A Guidebook on the Hydrologic Effects of Urban Land Use*. Circular 554. Washington, D.C.: U. S. Geological Survey, 18 pp.

Lin, Y.-L., S. Chiao, T.-A. Wang, M. L. Kaplan, and R. P. Weglarz. 2001. Some common ingredients for heavy orographic rainfall. *Weather and Forecasting*, **16**, 633-660.

Liu, H., V. Chandrasekar, and G. Xu. 2001. An adaptive neural network scheme for radar rainfall estimation from WSR-88D observations. *Journal of Applied Meteorology*, **40**, 2038-2050.

Lowndes, S. 1968. Forecasting large 24-h rainfall totals in the Dee and Clwyd River Authority Area from September to February. *Meteorological Magazine*, **97**, 226-235.

Maddox, R. A., L. R. Hoxit, C. F. Chappell, and F. Caracena. 1978. Comparison of meteorological aspects of the Big Thompson and Rapid City flash floods. *Monthly Weather Review*, **106**, 375-389.

Maddox, R. A., C. F. Chappell, and L. R. Hoxit. 1979. Synoptic and mesoscale aspects of flash flood events. *Bulletin of the American Meteorological Society*, **60**, 115-123.

Maddox, R. A., F. Canova, and L. R. Hoxit. 1980. Meteorological conditions of flash flood events over the western United States. *Monthly Weather Review*, **108**, 1866-1877.

Maddox, R. A., J. Zhang, J. J. Gourley, and K. W. Howard. 2002. Weather Radar Coverage over the Contiguous United States. *Weather and Forecasting*, **17**(4), 927-934.

Manobianco, J., G. E. Taylor, and J. W. Zack. 1996. Workstation-based real-time mesoscale modeling designed for weather support to operations at the Kennedy Space Center and Cape Canaveral Air Station. *Bulletin of the American Meteorological Society*, **77**, 653-672.

Marwitz, J. D. 1983. The kinematics of orographic airflow during Sierra storms. *Journal of the Atmospheric Sciences*, **40**, 1218-1227.

Mass, C. 2003. IFPS and the Future of the National Weather Service. *Weather and Forecasting*, **18**, 687-691.

Mass, C. F., D. Ovens, K. Westrick, and B. A. Colle. 2002. Does increasing horizontal resolution produce more skillful forecasts? *Bulletin of the American Meteorological Society*, **83**, 407-430.

Masutani, M., and A. Leetmaa. 1999. Dynamical mechanisms of the 1995 California floods. *Journal of Climate*, **12**, 3220-3236.

Matejka, T. J., R. A. Houze, and P. V. Hobbs. 1980. Microphysics and dynamics of clouds associated with mesoscale rainbands in extratropical cyclones. *Quarterly Journal of the Royal Meteorological Society*, **106**, 29-56.

Matrosov, S. Y., K. A. Clark, B. E. Martner, and A. Tokay. 2002. X-band polarimetric radar measurements of rainfall. *Journal of Applied Meteorology*, **41**, 941-952.

McNeill, D., and P. Freiberger. 1993. *Fuzzy Logic: The Revolutionary Computer Technology that is Changing Our World*. New York: Simon and Schuster, 319 pp.

Mecklenburg, S., V. A. Bell, D. S. Carrington, A. M. Cooper, R. J. Moore, and C. E. Pierce. 2002. Performance of radar-based rainfall forecasts on different spatial and temporal scales. *Abstracts, Vol. 2, International Conference on Quantitative Precipitation Forecasting*, Royal Meteorological Society, 120 pp.

Medina, S., and R. A. Houze, Jr. 2003. Air motions and precipitation growth in Alpine storms. *Quarterly Journal of the Royal Meteorological Society*, **129**, 345-371.

Megenhardt, D., C. K. Mueller, N. Rehak, and G. Cunning. 2000. Evaluation of the National Convective Weather Forecast product. Pp. 171-176 in *Preprints, 9th Conference on Aviation, Range and Aerospace Meteorology*, American Meteorological Society.

Megenhardt, D., C. K. Mueller, S. Trier, D. Ahijevych, N. Rehak, and G. Cunning. 2004. Description of National Convective Weather Forecast (NCWF-2). *Preprints, 11th Conference on Aviation, Range and Aerospace Meteorology*, American Meteorological Society.

Meiggs, R. K., M. R. Peroutka, D. A. Ruth. 1998. Implementing Interactive Forecast Preparation Nationwide. Pp. 359-363 in *Preprints 14th International Conference on Interactive Information and Processing Systems for Meteorology, Oceanography, and Hydrology*, American Meteorological Society.

Metcalf, J. I., and J. S. Ussailis. 1984. Radar system errors in polarization diversity measurements. *Journal of Atmospheric and Oceanic Technology*, **1,** 105-114.

Michalakes, J., S. Chen, J. Dudhia, L. Hart, J. Klemp, J. Middlecoff, and W. Skamarock. 2001. Development of a next generation regional weather research forecast model. Pp. 269-276 in *Developments in Teracomputing: Proceedings of the Ninth ECMWF Workshop on the Use of High Performance Computing in Meteorology,* W. Zwieflhofer and N. Kreitz, eds. Singapore: World Scientific.

Miller, N. L., and J. Kim. 1996. Numerical prediction of precipitation and river flow over the Russian River watershed during the January 1995 storms. *Bulletin of the American Meteorological Society*, **77**, 101-105.

Mitchell, T. P., and W. Blier. 1997. The variability of wintertime precipitation in the region of California. *Journal of Climate,* **10,** 2261-2276.

Mogil, H. M., J. C. Monro, and H. S. Groper. 1978. NWS's flash flood warning and disaster preparedness programs. *Bulletin of the American Meteorological Society,* **59,** 690-699.

Monteverdi, J., and J. Null. 1997. El Niño and California precipitation. Western Region Technical Attachment 97-37. Available online at *http://tornado.sfsu.edu/geosciences/elnino/elnino.html.* Accessed September 30, 2004.

Morin, E., W. F. Krajewski, D. C. Goodrich, X. Gao, and S. Sorooshian. 2003. Estimating rainfall intensities from weather radar data: The scale-dependency problem. *Journal of Hydrometeorology*, **4**, 782-797.

Moszkowicz, S., G. J. Ciach, and W. F. Krajewski. 1994. Statistical detection of anomalous propagation in radar reflectivity patterns. *Journal of Atmospheric and Oceanic Technology,* **11,** 1026-1034.

Mueller, C., T. Saxen, R. Roberts, J. Wilson, T. Betancourt, S. Dettling, N. Oien, and J. Yee. 2003. NCAR Auto-Nowcast System. *Weather and Forecasting*, **18**, 545-561.

Mullen, S. L., and R. Buizza, 2001. Quantitative precipitation forecasts over the United States by the ECMWF ensemble prediction system. *Monthly Weather Review,* **129,** 638-663.

Murphy, A. H. 1977. The value of climatological, categorical and probabilistic forecasts in the cost-loss ratio situation. *Monthly Weather Review.* **105**(7), 803-816.

NCDC (National Climatic Data Center). 1982. *Storm Data,* **24,** No. 1. 8-9.

NCDC. 1995. *Storm Data*, **37**, 111 pp.

NCDC. 1998. *Storm Data*, **40**, 188 pp.

Neiman, P. J., F. M. Ralph, A. B. White, D. E. Kingsmill, and P. O. G. Persson. 2002. The statistical relationship between upslope flow and rainfall in California's coastal mountains: Observations during CALJET. *Monthly Weather Review*, **130,** 1468-1492.

Neiman, P. J., P. O. G. Persson, F. M. Ralph, D. P. Jorgensen, A. B. White, and D. E. Kingsmill. 2004. Modification of fronts and precipitation by coastal blocking during an intense landfalling winter storm in southern California: Observations during CALJET. *Monthly Weather Review*, **132**, 242-273.

Nesbitt, S. W., E. J. Zipser, and D. J. Cecil. 2000. A census of precipitation features in the tropics using TRMM: radar, ice scattering, and lightning observations. *Journal of Climate,* **13**, 4087-4106.

NEXRAD Joint System Program Office, 1983. *NEXRAD Siting Handbook*. Silver Spring, Md.: NEXRAD Joint System Program Office.

Nicholas, A. P., and D. E. Walling. 1997. Modelling flood hydraulics and overbank deposition on river floodplains. *Earth Surface Processes and Landforms,* **22,** 59-77.

Nijssen. B., and D. P. Lettenmaier. 2004. Effect of precipitation sampling error on simulated hydrological fluxes and states: Anticipating the Global Precipitation Measurement satellites. *Journal of Geophysical Research,* **109D,** DOI: 10.1029/2003JD003497.

NRC (National Research Council). 1995. *Assessment of NEXRAD Coverage and Associated Weather Services*. Washington, D.C.: National Academy Press.

NRC. 2002. *Weather Radar Technology Beyond NEXRAD*. Washington, D.C.: National Academy Press.

NRC. 2003. *Fair Weather: Effective Partnerships in Weather and Climate Services*. Washington D.C.: The National Academies Press.

NRC. 2004. *Assessing the National Streamflow Information Program*. Washington, D.C.: The National Academies Press.

Null, J. 2001. El Niño and La Niña ... their relationship to California flood damage. Available online at *http://ggweather.com/nino/calif_flood.html*. Accessed August 17, 2004.

NWS (National Weather Service). 1997. *Automated Local Flood Warning Systems Handbook*. Weather Service Hydrology Handbook No. 2. Washington, D.C.: U.S. Government Printing Office.

NWS. 2001a. *Sulphur Mountain Doppler Radar: A Performance Study*. NOAA Technical Memorandum WB-267. Washington, D.C.: National Weather Service, 65 pp.

NWS. 2001b. Implementation of Advanced Hydrologic Prediction Service. Available online at *http://www.nws.noaa.gov/oh/ahps/Implementation_Plan.pdf*. Accessed August 24, 2004.

NWS. 2002a. Weather Surveillance Radar – 1988 Doppler (WSR-88D) Integrated Logistics Support Plan, R400-IS301C, 15 March 2002. Available online at *http://www.roc.noaa.gov/ilspfinal.pdf*. Accessed September 30, 2004.

NWS. 2002b. Advanced Hydrologic Prediction Services: Concepts of Services and Operations. Available online at *http://www.nws.noaa.gov/om/water/AHPSconcept.pdf*. Accessed August 24, 2004.

NWS. 2003a. Working together to save lives: National Weather Service strategic plan for FY 2003-FY 2008. Available online at *http://www.spo.noaa.gov/pdfs/LO%20Strat%20Plans%202003%20Final/newnwssp%2010-01-03%20final.pdf*. Accessed June 8, 2004.

NWS. 2003b. National Weather Service policy directive 10-16. January 6, 2003. Operations and Service: Performance. Available online at *http://www.nws.noaa.gov/directives/010/pd01016a.pdf*. Accessed June 4, 2004.

NWS. 2004a. Turn Around, Don't Drown. Available online at *http://www.srh.noaa.gov/tadd/*. Accessed June 19, 2004.

NWS. 2004b. National Weather Service Hydrology Statistics from 1/1/96 through 12/31/03. Data provided by the National Weather Service in February 2004.

NWS. 2004c. National Weather Service Hydrologic Glossary. Available online at *http://www.crh.noaa.gov/hsd/hydefa-c.html*. Accessed May 14, 2004.

NWS. 2004d. National Weather Service Instruction 10-1601. July 20, 2004. Operations and Service Performance, NWSPD 10-16. Verification procedures. Available online at *http://www.nws.noaa.gov/directives/010/pd01016001b.pdf*. Accessed September 30, 2004.

O'Connor, J. E., and J. E. Costa. 2004. Spatial distribution of the largest rainfall-runoff floods from basins between 2.6 and 26000 km^2 in the United States and Puerto Rico. *Water Resources Research*, **40,** W01107, doi:10.1029/2003WR002247.

Ogden, F. L., H. O. Sharif, S. U. S. Senarath, J. A. Smith, M. L. Baeck, and J. R. Richardson. 2000. Hydrologic analysis of the Fort Collins, Colorado flash flood of 1997. *Journal of Hydrology,* **228,** 82-100.

Pandey, G. R., D. R. Cayan, and K. P. Georgakakos. 1999. Precipitation structure in the Sierra Nevada of California during winter. *Journal of Geophysical Research,* **104**, 12019-12030.

Pappert, R. A., and C. L. Goodhart. 1977. Case studies of beyond-the-horizon propagation in tropospheric ducting environments. *Radio Science*, **12**, 75-87.

Paris, J. F. 1997a. *A Critique of the Sulphur Mountain, CA NexRad (WSR-88D) Site*. Seaside, Calif.: Spatial Information, Visualization, and Analysis Resources Center, California State University, 29 pp.

Paris, J. F. 1997b. *A Critique of the Sulphur Mountain, CA NexRad (WSR-88D) Site, Addendum I*. Seaside, Calif.: Spatial Information, Visualization, and Analysis Resources Center, California State University, 4 pp.

Paris, J. F. 1998. *A Critique of the Sulphur Mountain, CA NexRad (WSR-88D) Site, Addendum II*. Salinas, Calif.: Paris and Associates, 18 pp.

Paris, J. F. 2001. *Performance of the Sulphur Mountain NexRad vis-à-vis neighboring NexRads relative to the prediction of flash floods in Los Angeles County and Ventura County, California*. Seaside, Calif.: Spatial Information, Visualization, and Analysis Resources Center, California State University, 25 pp.

Petersen, W. A., L. D. Carey, S. A. Rutledge, J. C. Knievel, N. J. Doeskin, R. H. Johnson, T. B. McKee, T. Vonder Haar, and J. F. Weaver. 1999. Mesoscale and radar observations of the Fort Collins flash flood of 28 July 1997. *Bulletin of the American Meteorological Society,* **80,** 191-216.

Polger, P. D., B. S. Goldsmith, R. C. Przywarty, and J. R. Bocchieri. 1994. National Weather Service warning performance based on the WSR-88D. *Bulletin of the American Meteorological Society*, **75**, 203-214.

Potter, K. W. 1991. Hydrological impacts of changing land management practices in a moderate-sized agricultural catchment. *Water Resources Research*, **27**, 845-855.

Pratte, F., R. J. Keeler, R. Gagnon, and D. Sirmans. 1995. Clutter processing during anomalous propagation conditions. Pp. 139-141 in *Preprints, 27th Conference on Radar Meteorology,* American Meteorological Society.

Puzzo, D. C., S. W. Troxel, M. A. Meister, M. E. Weber, and J. V. Pieronek. 1989. Appendix I: Methods for filtering clutter as a result of anomalous propagation. In *ASR-9 Weather Channel Test Report.* Project Report ATC-165. Lexington: Lincoln Laboratory, Massachusetts Institute of Technology.

QJRMS (Quarterly Journal of the Royal Meteorological Society). 2003. **129**(588).

Radar Operations Center. 2002. *Interface Control Document for the RPG to Class 1 User.* Document 2620001C. Norman, Okla.: Radar Operations Center.

Ralph. F. M., P. J. Neiman, D. E. Kingsmill, P. O. G. Persson, A. B. White, E. T. Strem, E. D. Andrews, and R. C. Antweiler. 2003. The impact of a prominent rain shadow on flooding in California's Santa Cruz mountains: A CALJET case study and sensitivity of the ENSO cycle. *Journal of Hydrometeorology*, **4**, 1243-1264.

Ralph, F. M., P. J. Neiman, and G. A. Wick. 2004. Satellite and CALJET aircraft observations of atmospheric rivers over the eastern North Pacific Ocean during the winter of 1997/98. *Monthly Weather Review*, **132**, 1721-1745.

Reinking, R. F., and J. F. Boatman. 1986. Upslope precipitation events. Pp. 437-471 in *Mesoscale Meteorology and Forecasting*, P. S. Ray, ed. Boston: American Meteorological Society.

Reynolds, D. W., and A. S. Dennis. 1986. Review of the Sierra cooperative pilot project. *Bulletin of the American Meteorological Society*, **67**, 513-523.

Reynolds, D. W., and A. P. Kuciauskas. 1988. Remote and in situ observations of Sierra Nevada winter mountain clouds: Relationships between mesoscale structure, precipitation and liquid water. *Journal of Applied Meteorology*, **27**, 140-156.

Rich, S. T. 1992. Integrating wind profiler data into forecast and warning operations at NWS field offices. NOAA Tech. Memo. NWS SR-141, 34. Boulder, Colo.: NOAA Forecast Systems Lab.

Richardson, D. S. 2000. Skill and Economic Value of the ECMWF Ensemble Prediction System. *Quarterly Journal of the Royal Meteorological Society*, **126**, 649-668.

Roberts, R. D., and S. Rutledge. 2003. Nowcasting storm initiation and growth using GOES-8 and WSR-88D data. *Weather and Forecasting*, **18**, 562-584.

Roberts, R., T. Saxen, C. Mueller, and J. Wilson. 2003. A forecaster-computer interactive capability of the NCAR Auto-Nowcaster system for improved storm initiation nowcasts. Pp. J20-J23 in *Preprints, 31st Conference on Radar Meteorology*, American Meteorological Society.

Rodriguez-Iturbe, I., and A. Rinaldo. 1997. *Fractal River Basins: Chance and Self-Organization*. New York: Cambridge University Press, 547 pp.

Rogers, R. R. 1979. *A Short Course in Cloud Physics*. 2nd edition. New York: Pergamon Press, 235 pp.

Romanowicz, R., and K. Beven. 2003. Estimation of flood inundation probabilities as conditioned on event inundation maps. *Water Resources Research*, **39**, 1073, doi:10.1029/2001WR001056.

Rose Institute. 1997. *The National Weather Service's Tower in the Upper Ojai: A Case History*. Claremont, Calif.: Rose Institute of State and Local Government.

Rose Institute. 1998. *The National Weather Service's NEXRAD Radar Tower at Sulphur Mountain in the Upper Ojai Valley*. Claremont, Calif.: Rose Institute of State and Local Government.

Rotunno, R., and R. Ferretti. 2001. Mechanisms of intense Alpine rainfall. *Journal of the Atmospheric Sciences*, **58**, 1732-1749.

Ruth, D. A., M. A. Mathewson, T. J. LeFebvre, and P. Wu. 1998. Interpretation and editing techniques for interactive forecast preparation. Pp. 345-349 in *Preprints, 14th International Conference on Interactive Information and Processing Systems for Meteorology, Oceanography, and Hydrology,* American Meteorological Society.

Ruth, D. P. 2002. Interactive forecast preparation—The future has come. Pp. 20-22 in *Preprints, Interactive Symposium on the Advanced Weather Interactive Processing System (AWIPS)*, American Meteorological Society.

Saffle, R. E., M. I. Istok and L. D. Johnson. 2001. NEXRAD Open Systems—Progress and Plans. Pp. 690-693 in *Preprints, 30th International Conference on Radar Meteorology*, American Meteorological Society.

San Bernardino County Public Information Office. 2004. County continues to alert public to post-fire flood dangers. San Bernadino County news press release, January 3, 2004. Available online at *http://www.ci.loma-linda.ca.us/PDF-Files/flood%20control%20release%201-12-04.pdf*. Accessed October 1, 2004.

Saxen, T., C. Mueller, J. Wilson, R. Roberts, E. Nelson, D. Ahijevych, and S. Trier. 2004. Updates to the Auto-Nowcaster for the 2004 convective weather season. *Preprints, 11th Conference on Aviation, Range and Aerospace Meteorology*, American Meteorological Society.

Schwarz, F. K. 1970. The unprecedented rains associated with the remnants of hurricane Camille. *Monthly Weather Review*, **98**, 851-859.

Seed, A. W. 2003. A dynamic and spatial scaling approach to advection forecasting. *Journal of Applied Meteorology*, **42**, 381-388.

Seguin, W. R. 2002. AWIPS—An end-to-end look. Pp. J47-J51 in *Preprints, 18th International Conference on Interactive Information and Processing Systems (IIPS) for Meteorology, Oceanography, and Hydrology*, American Meteorological Society.

Sekhon, R. S., and R. C. Srivastava. 1971. Doppler radar observations of drop size distributions in a thunderstorm. *Journal of the Atmospheric Sciences*, **28**, 983-994.

Seliga, T. A., and V. N. Bringi. 1976. Potential use of radar differential reflectivity measurements at orthogonal polarizations for measuring precipitation. *Journal of Applied Meteorology*, **15**, 69-76.

Seo, D.-J., J. Breidenbach, R. Fulton, D. Miller, and T. O'Bannon. 2000. Real-time adjustment of range-dependent biases in WSR-88D rainfall estimates due to nonuniform vertical profile of reflectivity. *Journal of Hydrometeorology*, **1**, 222-240.

Seo, D.-J., V. Koren, and N. Cajina. 2003. Real-time variational assimilation of hydrologic and hydrometeorological data into operational hydrologic forecasting. *Journal of Hydrometeorology*, **4**, 627-641.

Serafin, R. J., and J. W. Wilson. 2000. Operational weather radar in the United States. Progress and opportunity. *Bulletin of the American Meteorological Society,* **81,** 501-518.

Sharif, H. O., F. L. Ogden, W. F. Krajewski, and M. Xue. 2004. Statistical analysis of radar rainfall error propagation. *Journal of Hydrometeorology*, **5**, 199-212.

Shiono, K., Y. Muto, D. W. Knight, and A. F. L. Hyde. 1999. Energy losses due to secondary flow and turbulence in meandering channels with overbank flows. *Journal of Hydraulic Research*, **37**, 641-664.

Sinclair, M. R. 1994. A diagnostic model for estimating orographic precipitation. *Journal of Applied Meteorology*, **33**, 1163-1175.

Sivapragasam, C., S. Y. Liong, and M. F. K. Pasha. 2001. Rainfall and runoff forecasting with SSA-SVM approach. *Journal of Hydroinformatics*, **3**, 141-152.

Skolnik, M. I. 1980. *Introduction to Radar Systems*. Singapore: McGraw-Hill International Editions. 581 pp.

Smith, J. A., M. L. Baeck, M. Steiner, and A. J. Miller. 1996. Catastrophic rainfall from an upslope thunderstorm in the central Appalachians: The Rapidan storm of June 27, 1995. *Water Resources Research*, **32**, 3099-3113.

Smith, J. A., M. L. Baeck, J. E. Morrison, and P. Sturdevant-Rees. 2000. Catastrophic rainfall and flooding in Texas. *Journal of Hydrometeorology*, **1**, 5-25.

Smith, J. A., M. L. Baeck, Y. Zhang, and C. A. Doswell III. 2001. Extreme rainfall and flooding from supercell thunderstorms. *Journal of Hydrometeorology*, **2**, 469-489.

Smith, J. A., M. L. Baeck, J. E. Morrison, P. Sturdevant-Rees, D. F. Turner-Gillespie, and P. D. Bates. 2002. The regional hydrology of extreme floods in an urbanizing drainage basin. *Journal of Hydrometeorology*, **3**, 267-282.

Smith, R. B. 1979. The influence of mountains on the atmosphere. *Advances in Geophysics*, **21**, 87-230.

Smith, R., J. Paegle, T. Clark, W. Cotton, D. Durran, G. Forbes, J. Marwitz, C. Mass, J. McGinley, H. Pan, and F.M. Ralph. 1997. Local and remote effects of mountains on weather: Research needs and opportunities. *Bulletin of the American Meteorological Society*, **78**, 877-892.

Snyder, C. and F. Zhang. 2003. Assimilation of simulated Doppler radar observations with an ensemble Kalman filter. *Monthly Weather Review*, **131**, 1663-1667.

Steiner, M., and J. A. Smith. 2002. Use of three-dimensional reflectivity structure for automated detection and removal of nonprecipitating echoes in radar data. *Journal of Atmospheric and Oceanic Technology*, **19**, 673-686.

Steiner, M., J. A. Smith, and R. Uijlenhoet. 2004. A microphysical interpretation of radar reflectivity-rain rate relationships. *Journal of the Atmospheric Sciences*, **61**, 1114-1131.

Steiner, M., J. A. Smith, S. J. Burges, C. V. Alonso, and R. W. Darden. 1999. Effect of bias adjustment and rain gauge data quality control on radar rainfall estimation. *Water Resources Research*, **35**, 2487-2503.

Stensrud D. J., H. E. Brooks, J. Du, M. S. Tracton, and E. Rogers. 1999. Using Ensembles for Short-Range Forecasting, *Monthly Weather Review*, **127**, 433-446.

Stoelinga, M. T., P. V. Hobbs, C. F. Mass, J. D. Locatelli, N. A. Bond, B. A. Colle, R. A. Houze, Jr., and A. Rangno. 2003. Improvement of microphysical parameterization through observational verification experiment (IMPROVE). *Bulletin of the American Meteorological Society*, **84**, 1807-1826.

Storm Prediction Center. 2004. The online tornado FAQ: Tornado climatology. Available online at *http://www.spc.noaa.gov/faq/tornado/#Climatology*. Accessed June 19, 2004.

Straka, J. M., D. S. Zrnic, and A. V. Ryzhkov. 2000. Bulk hydrometeor classification and quantification using polarimetric radar data: Synthesis of relations. *Journal of Applied Meteorology*, **39**, 1341-1372.

Sturdevant-Rees, P., J. A. Smith, J. Morrison, and M. L. Baeck. 2001. Tropical storms and the flood hydrology of the central Appalachians. *Water Resources Research*, **37**, 2143-2168.

Sun, J. 2004. Numerical prediction of thunderstorms: Fourteen years later. In *Atmospheric Turbulence and Mesoscale Meteorology*, E. Fedorovich, R. Rotunno, and B. Stevens, eds. Cambridge, U.K.: Cambridge University Press.

Sun, J., and N. A. Crook. 1997. Dynamical and microphysical retrieval from Doppler radar observations using a cloud model and its adjoint. Part I: Model development and simulated data experiments. *Journal of the Atmospheric Sciences*, **54**, 1642-1661.

Sun, J., and N. A. Crook. 1998. Dynamical and microphysical retrieval from Doppler radar observations using a cloud model and its adjoint. Part II: Retrieval experiments of an observed Florida convective storm. *Journal of the Atmospheric Sciences*, 55, 835-852.

Sun, J., and N. A. Crook. 2001. Real-time low-level wind and temperature analysis using single WSR-88D data. *Weather and Forecasting*, **16**, 117-132.

Testud, J., S. Oury, P. Amayenc, and R. A. Black. 2001. The concept of "normalized" distributions to describe raindrop spectra: a tool for cloud physics and cloud remote sensing. *Journal of Applied Meteorology*, **40**, 1118-1140.

Thompson, R. A. 2001. Flash flood event of 6 February 1998: A case study. Western Region Technical Attachment 01-08. Available online at *http://www.wrh.noaa.gov/wrhq/01TAs/0108/*. Accessed October 1, 2004.

Toth, Z., O. Talagrand, G. Candille, and Y. Zhu. 2003. Probability and ensemble forecasts. Pp. 137-164 in *Environmental Forecast Verification: A Practitioner's Guide in Atmospheric Science*. I. T. Jolliffe and D. B. Stephenson, eds. New York: John Wiley and Sons.

Tracton M. S., and J. Du. 2001. Application of the NCEP/EMC short-range ensemble forecast system (SREF) to predicting extreme precipitation events. Pp. 64-65 in *Preprints, Symposium on Precipitation Extremes: Prediction, Impacts, and Responses*, American Meteorological Society.

Turner-Gillespie, D. F., J. A. Smith, and P. D. Bates. 2003. Attenuating reaches and the regional flood response of an urbanizing drainage basin. *Advances in Water Resources*, **26**, 673-684.

VCWPD (Ventura County Watershed Protection District). 2004. Available online at *http://publicworks.countyofventura.org/fc/Index.htm*. Accessed June 25, 2004.

Verlinde, J., and W. R. Cotton. 1993. Fitting microphysical observations of non-steady convective clouds to a numerical model: An application of the adjoint technique of data assimilation to a kinematic model. *Monthly Weather Review*, **121**, 2776-2793.

Vignal, B., G. Galli, J. Joss, U. Germann. 2000. Three methods to determine profiles of reflectivity from volumetric radar data to correct precipitation estimates. *Journal of Applied Meteorology*, **39**, 1715-1726.

Vivekanandan, J., D. N. Yates, and E. A. Brandes. 1999. The influence of terrain on rainfall estimates from radar reflectivity and specific propagation phase observations. *Journal of Atmospheric and Oceanic Technology*, **16**, 837-845.

Walser, A., D. Lüthi, and C. Schär. 2004. Predictability of precipitation in a cloud-resolving model. *Monthly Weather Review*, **132**, 560-577.

Warner, T. T., E. A. Brandes, J. Z. Sun, D. N. Yates, and C. K. Mueller. 2000. Prediction of a flash flood in complex terrain. Part I: a comparison of rainfall estimates from radar, and very short range rainfall simulations from a dynamic model and an automated algorithmic system. *Journal of Applied Meteorology*, **39**, 797-814.

Weber, M. E. 2000. FAA Surveillance radar data as a complement to the WSR-88D network. *Preprints, 9th Conference on Aviation Range and Aerospace Meteorology*, American Meteorological Society.

Weber, M. E., M. L. Stone, and J. A. Cullen. 1993. Anomalous propagation associated with thunderstorm outflows. Pp. 238-240 in *Preprints, 26th International Conference on Radar Meteorology*, American Meteorological Society.

Westrick, K. J., and C. F. Mass. 2001. An evaluation of a high-resolution hydrometeorological modeling system for prediction of a cool-season flood event in a coastal mountainous watershed. *Journal of Hydrometeorology*, **2**, 161-180.

Westrick, K. J., C. F. Mass, and B. A. Colle. 1999. The limitations of the WSR-88D radar network for quantitative precipitation measurement over the coastal western United States. *Bulletin of the American Meteorological Society*, **80**(11), 2289-2298.

Westrick, K. J., P. Storck, and C. F. Mass. 2002. Description and evaluation of a hydrometeorological forecast system for mountainous watersheds. *Weather and Forecasting*, **17**, 250-262.

Weygandt, S., A. Shapiro, and K. Droegemier. 2002a. Retrieval of model initial fields from single Doppler observations of a supercell thunderstorm. Part I. Single Doppler velocity retrieval. *Monthly Weather Review*, **130**, 433-453.

Weygandt, S., A. Shapiro, and K. Droegemier. 2002b. Retrieval of model initial fields from single Doppler observations of a supercell thunderstorm. Part II. Thermodynamic retrieval and numerical prediction. *Monthly Weather Review*, **130**, 454-476.

White, A. B., P. J. Neiman, F. M. Ralph, D. E. Kingsmill, and P. O. G. Persson. 2003. Coastal orographic rainfall processes observed by radar during the California land-falling jets experiment. *Journal of Hydrometeorology*, **4**, 264-282.

Willis, P. T. 1984. Functional fits to some observed drop size distributions and parameterization of rain. *Journal of the Atmospheric Sciences*, **41**, 1648-1661.

Wilson, J. W., and E. A. Brandes. 1979. Radar measurement of rainfall—A summary. *Bulletin of the American Meteorological Society*, **60**(9), 1048-1060.

Wilson J. W., and W. E. Schreiber. 1986. Initiation of convective storms at radar-observed boundary-layer convergence lines. *Monthly Weather Review*, **114**, 2516-2536.

Wilson, J. W., and D. M. Pollock. 1974. Rainfall measurements during hurricane Agnes by three overlapping radars. *Journal of Applied Meteorology*, **13**, 835-844.

Wilson, J. W., R. D. Roberts, C. Kessinger, and J. McCarthy. 1984. Microburst wind structure and evaluation of Doppler radar for airport wind shear detection. *Journal of Applied Meteorology*, **23**, 898-915.

Winchel, M., H. V. Gupta, and S. Sorooshian. 1998. On the simulation of infiltration- and saturation-excess runoff using radar-based rainfall estimates: Effects of algorithm uncertainty and pixel aggregation. *Water Resources Research*, **34**, 2655-2670.

Wolff, G. C., and S. J. Burges. 1994. An analysis of the influence of river channel properties on flood frequency. *Journal of Hydrology*, **153**, 317-337.

Wolfson, M. M., G. E. Forman, R. G. Hallowell, and M. P. Moore. 1998. The growth and decay tracker. Pp. 58-62 in *Preprints, Eighth Conference on Aviation, Range and Aerospace Meteorology*, American Meteorological Society.

Woltemade, C. J., and K. W. Potter. 1994. A watershed modeling analysis of fluvial geomorphic influences on flood peak attenuation. *Water Resources Research*, **30**, 1933-1942.

Wood V. T., R. A Brown, and S. V. Vasiloff. 2003a. Improved detection using negative elevation angles for mountaintop WSR-88Ds. Part II: Simulations of the three radars covering Utah. *Weather and Forecasting*, **18**(3), 393-403.

Wood, V. T., R. A. Brown, S. V. Vasiloff. 2003b. Negative elevation angles for mountaintop radar. *Bulletin of the American Meteorological Society*, **84**, 996-997.

Xiao, Q., Y.-H. Kuo, J. Sun, W.-C. Lee, E. Lim, Y. Guo, and D. M. Barker. 2004. Assimilation of Doppler radar observations with a regional 3D-Var system: Impact of Doppler velocities on forecasts of a heavy rainfall case. *Journal of Applied Meteorology* (submitted).

Xu, M., D. J. Stensrud, J.-W. Bao, and T. T. Warner. 2001. Applications of the adjoint technique to short-range ensemble forecasting of mesoscale convective systems. *Monthly Weather Review*, **129**, 1395-1418.

Xu, M., N. A. Crook, Y. Liu, and R. Rasmussen. 2003. Radar data assimilation for real-time short-term forecasting of snowbands. Pp. 1024-1027 in *Preprints, 31st Conference on Radar Meteorology*, American Meteorological Society.

Yates, D. N., T. T. Warner, G. H. Leavesley. 2000. Prediction of a flash flood in complex terrain. Part II. A comparison of flood discharge simulations using rainfall input from radar, a dynamic model, and an automated algorithmic system. *Journal of Applied Meteorology*, **39**, 815-825.

Yates, D., T. T. Warner, E. A. Brandes, G. H. Leavesley, J. Sun, and C. K. Mueller. 2001. Evaluation of flash-flood discharge forecasts in complex terrain using precipitation. *Journal of Hydrologic Engineering*, **6**, 265-274.

Young, C. B., B. R. Nelson, A. A. Bradley, J. A. Smith, C. D. Peters-Lidard, A. Kruger, and M. L. Baeck. 1999. An evaluation of NEXRAD precipitation estimates in complex terrain. *Journal of Geophysical Research-Atmospheres*, **104**, 19691-19703.

Zawadzki, I. 1984. Factors affecting the precision of radar measurements of rain. Pp. 251-256 in *Preprints, 22nd Conference on Radar Meteorology*, American Meteorological Society.

Zeng, Z., S. E. Yuter, R. A. Houze, and D. E. Kingsmill. 2001. Microphysics of the rapid development of heavy convective precipitation. *Monthly Weather Review*, **129**, 1882-1904.

Zhang, F., C. Snyder, and J. Sun. 2004. Impacts of initial estimate and observations on convective-scale data assimilation with an ensemble Kalman filter. *Monthly Weather Review*, **132**, 1238-1253.

Zhang, Y., J. A. Smith, and M. L. Baeck. 2001. The hydrology and hydrometeorology of extreme floods in the Great Plains of eastern Nebraska. *Advances in Water Resources*, **24**, 1037-1049.

Zrnic, D. S., and A. V. Ryzhkov. 1999. Polarimetry for Weather Surveillance Radars. *Bulletin of the American Meteorological Society*, **80**(3), 389-406.

Zrnic, D. S., T. D. Keenan, L. D. Carey, and P. May. 2000. Sensitivity analysis of polarimetric variables at a 5-cm wavelength in rain. *Journal of Applied Meteorology*, **39**, 1514-1526.

Appendixes

Appendix A

Origin of the Study

LANGUAGE FROM THE 2003 CONSOLIDATED APPROPRIATIONS RESOLUTION, PUBLIC LAW 108-7

The conferees direct NOAA to commission the National Academy of Sciences to conduct a study to assess the availability, performance, and capability of the NWS NEXRAD located on Sulphur Mountain in Ventura County, California, to detect heavy precipitation and aid forecasters at the Los Angeles Weather Forecast Office in providing flash flood warnings and forecasts, and on the basis of that study, to provide the Under Secretary of Commerce for Oceans and Atmosphere with a report on the performance of that mission by the NWS. The report also should include any recommendations for improving the accuracy and timeliness of flash flood warnings in and around western Los Angeles and Ventura Counties, California.

LETTER FROM U.S. SENATOR BARBARA BOXER

BARBARA BOXER
CALIFORNIA

COMMITTEES:
COMMERCE, SCIENCE, AND TRANSPORTATION
ENVIRONMENT AND PUBLIC WORKS
FOREIGN RELATIONS

United States Senate

HART SENATE OFFICE BUILDING
SUITE 112
WASHINGTON, DC 20510-0505
(202) 224-3553
http://boxer.senate.gov/contact

APR 15 2003

April 7, 2003

Dr. Bruce Alberts
President
National Academy of Sciences
500 5th St. NW
Washington, DC 20001

Dear Dr. Alberts:

I am writing regarding the National Academy of Sciences (NAS) study on the performance of the Sulphur Mountain National Weather Service NEXRAD, which was required as part of the conference report to the Fiscal Year 2003 omnibus appropriations bill (P.L. 108-7).

I requested that this study be conducted because of the possible threat that flash floods pose to public safety in the heavily populated areas of Los Angeles and Ventura counties.

Concerns have been raised that the NEXRAD is not effective in providing sufficient warnings. Although the General Accounting Office studied the performance of the Sulphur Mountain radar in 1998, concerns have been raised that the study contained conflicting evidence, and the conclusions did not always appear to match the data.

As you conduct this study, I hope that you will specifically examine the warning failure rate and radar data gap. Attached is a paper further detailing these concerns.

Thank you for your careful evaluation, and I look forward to the NAS's final recommendations.

Sincerely,

Barbara Boxer
United States Senator

1700 MONTGOMERY STREET, SUITE 240, SAN FRANCISCO, CA 94111, (415) 403-0100
312 NORTH SPRING STREET, SUITE 1748, LOS ANGELES, CA 90012, (213) 894-5000
501 'I' STREET, SUITE 7-600, SACRAMENTO, CA 95814, (916) 448-2787
1130 'O' STREET, SUITE 2450, FRESNO, CA 93721, (559) 497-5109
600 'B' STREET, SUITE 2240, SAN DIEGO, CA 92101, (619) 239-3884
201 NORTH 'E' STREET, SUITE 210, SAN BERNARDINO, CA 92401, (909) 888-8525

PRINTED ON RECYCLED PAPER

ADDITIONAL INFORMATION

• Since the Sulphur Mountain NEXRAD began operating and its data incorporated into forecasts by meteorologists in the Los Angeles Weather Forecasting Office, it has failed to warn of flash floods nearly 50 percent of the time. Of twelve flood events in the El Niño year of 1998, either no or insufficient warning time was given for five events.

• According to the administrator of the National Weather Service, the Sulphur Mountain NEXRAD is the only Doppler radar in the Southern California interlocking NEXRAD system whose mission is to detect low-level storms approaching the coast over water in the winter.

• A geo-spatial chart prepared by the Geo-Spatial Information Center at California State University, Monterey indicates a gap at 6,000 feet between the interlocking NEXRAD radar systems. At 5,000 feet this gap is considerably larger. The gap indicated by the chart has been created by the Sulphur Mountain NEXRAD beam being located too high and the reach of the Santa Ana Mountains' NEXRAD reach being very limited. The Santa Ana Mountains' NEXRAD barely extends beyond the coastline at the altitude of 5,000 feet. This gap creates a lack of available data.

• The preliminary NEXRAD site survey for the Los Angeles area stated that the Sulphur Mountain radar, in order to avoid anomalous propagations, needed to be placed at an elevation of over 2,000 feet. As a result, the Sulphur Mountain NEXRAD was placed at 2,750 feet. The height of inversion, where warmer air meets cooler air in the atmosphere, is 2,000 feet according to the site survey. Even though Sulphur Mountain NEXRAD is relied upon to provide continuous data to forecasters on the possibility of flash floods, the lower altitude of winter inversions was not taken into account in the site survey. In fact, the South Coast Air Quality Management District studied the years from 1950 to 1974, and found that during the month of January the coastal inversion exceeded 100 feet only three times.

Appendix B

Characteristics of the NEXRAD Radar

Parameter or Feature	Value or Description
Radar System	
Range of Observation	
Reflectivity	460 km
Velocity	230 km
Angular Coverage	
Azimuth	Full circle or sector
Elevation	Operational limits; –1° to +20°
Antenna	
Type	S-band, center-fed, parabolic dish
Reflector aperture	8.54-m (28-foot) diameter; circular
Beamwidth (one-way, 3 dB)	0.96° at 2.7 GHz; 0.88° at 3.0 GHz
Gain	45.8 dB at 2.85 GHz (midband)
Polarization	Linear horizontal
First side-lobe level	–29 dB
Steerability	360° azimuth; –1° to +45° elevation
Mechanical limits	–1° to +60°
Rotation rate	$30° s^{-1}$ (azimuth and elevation)
Angular acceleration	$15° s^{-2}$ (azimuth and elevation)
Pointing accuracy	±0.2°
Radome	
Type	Fiberglass skin foam sandwich
Diameter	11.89 m (39 feet.)
RF Loss (two-way)	0.3 ± 0.06 dB over 2.7–3.0-GHz band
Transmitter	
Type	Master Oscillator Power Amplifier (MOPA)
Frequency range	2.7–3.0 GHz
Peak power output (nominal)	500 kW into antenna
Pulsewidth (nominal)	1.57 μs (short pulse); 4.5 μs (long pulse) ± 4%
RF duty cycle (maximum)	0.002

Pulse Repetition Frequency	
Long pulse	322–422 Hz ± 1.7%
Short pulse	322–1282 Hz ± 1.7%
Waveform types	Contiguous and batch (phase-coded build 8.0)
Receiver	
Type	Linear
Tunability (frequency range)	2.7–3.0 GHz
Bandwidth (3 dB)	0.63 MHz (short pulse); 0.22 MHz (long pulse)
Phase control	Selectable
Receiver channels	Linear output I/Q
Dynamic range	95 dB max; 93 dB at 1-dB compression
Minimum detectable signal	–113 dBm (ORDA)
Noise temperature	450 K (ORDA)
Intermediate frequency	57.6 MHz
Sampling rate	600 kHz (This remains the same for the I/Q samples; IF samples will be at 72 MHz)
Signal Processor	
Type	Programmable
Parameters derived	Reflectivity; mean radial velocity; Doppler spectral width
Algorithms (respective)	Power averaging; pulse-pair; single-lag correlation
Accuracy (Standard Deviation)	
Reflectivity	<1 dB
Velocity and spectrum width	<1 m s^{-1}
Number of Pulses Averaged	
Reflectivity	6–64
Velocity and spectrum width	40–200
Range Resolution	
Reflectivity	1 km
Velocity and spectrum width	0.25 km
Azimuth Resolution	
Reflectivity	1°
Velocity and spectrum width	1°
Clutter canceller	Gaussian Model Adaptive Processing Filter, Spectral Domain
Clutter suppression	30–50 dB max
Radar Product Generator (RPG)	
RPG processor	64-bit reduced instruction set (RISC) digital computer
Shared memory	512-MB semiconductor memory
Wide-band communication	1.544 Mbit s^{-1} data rate
Narrow-band communication	Up to 13 of 14,400/4,800 bit s^{-1} 4-wire (1 may be 33,600 bit s^{-1} capable) 8 of 14,400/4,800 bit s^{-1} 2-wire Up to 9 Wide Area Network (WAN) connections
RPG Graphic Display Processor	
Principal user processor (PUP)	64-bit RISC digital computer

Communications	14,400/4,800 bit s^{-1}; 2- and 4-wire (maximum: 28 lines)
Video	Color, with split-screen and zoom features
Mass storage	Up to two 40-GB disks
Other	
National Climatic Data Center (NCDC)	National archive for NEXRAD data and other meteorological and climatological data.
Level II archive	Network interface located at the RPG. Digital base data are output from the Base Data Dissemination System (BDDS), which includes base reflectivity, mean radial velocity, and spectrum width. Data are sent electronically to the NCDC for permanent storage. (Completed nationwide in September 2004)
Level III archive	Interface is located at the RPG. A set of predetermined products defined in FMH-11 Part A is sent electronically to NCDC for permanent storage. (Completed nationwide in September 2004)
National Weather Radar Network	Consists of WSR-88D sites dispersed throughout the conterminous United States (CONUS) plus Department of Defense sites or non-CONUS Department of Transportation sites.

Appendix C

Chronology of the Siting of the Sulphur Mountain NEXRAD

Date	Event
August 1984-December 1984	SRI International, Inc. receives input on site suggestions from the Federal Aviation Administration Western Pacific Region, the National Weather Service Western Region, and the NEXRAD Joint System Program Office
March 1985	Analysis of the local atmospheric inversion statistics are provided to SRI International, Inc.
March-July 1985	SRI International, Inc. visits many sites in the Los Angeles area
April 1986	SRI International, Inc. completes the Preliminary Site Survey for the Los Angeles, California area (Sulphur Mountain site)
August 1987	SRI International, Inc. completes the In-Depth Site Survey for the Los Angeles, California area (Sulphur Mountain site)
1987	Sulphur Mountain selected as the best of 32 candidate locations
1992	SRI International, Inc. reviews alternative siting criteria
November 15, 1993	Construction of Sulphur Mountain NEXRAD begins
December 16, 1993	Sulphur Mountain NEXRAD installation complete
January 1994-March 1994	Sulphur Mountain NEXRAD tested
March 1994	Sulphur Mountain NEXRAD accepted
December 1994	Sulphur Mountain NEXRAD commissioned

Appendix D

National Weather Service Flash Flood Verification Procedures[1]

HYDROLOGIC VERIFICATION PROCEDURES FOR FLASH FLOOD WARNINGS

The Office of Climate Water and Weather Services (OCWWS) Performance Branch is responsible for the operation and maintenance of the automated flash flood warning (FFW) verification program.

Matching Warnings and Events

All warning data are automatically extracted from the warning products issued to the public. FFWs are issued by county. Since each county specified in a warning represents a separate verification area, a warning covering three counties is counted as three warned areas or three warnings. Events are automatically taken from the final Storm Data reports prepared by the WFOs. Storm Data reports entered as the event type "flash flood" verify an FFW.

For verification purposes, multiple flash flood events in the same county separated by less than 30 minutes are considered duplicates; therefore, only the first entry is made to the event database. This rule has the following exceptions:

[1]This information is excerpted from the National Weather Service Operations and Services Performance manual 10-1601 entitled "Verification Procedures" (NWS, 2004d). This manual was revised July 20, 2004, after the committee completed its analysis. The revised version includes quality assurance rules, which have been in effect since January 2002 but were inadvertently left out of the prior version, dated January 6, 2003. Because the committee adhered to the January 6, 2003, version, it did not utilize the quality assurance rules in its analysis. In particular, consistent application of Rule 2 would modify some of the entries in Table 7.1.

• any event that causes death or injury is included in the event database;

• any event that causes crop or property damage in excess of $500,000 is included in the event database;

• an event is not considered a duplicate if it is the only event verifying a warning.

Any event not recorded in the verification database due to the aforementioned duplicate rule still appears in the publication Storm Data.

Warnings and events are recorded in separate databases. Whenever an event occurs in a warned county, the following are recorded: one verified warning and one warned event. One unwarned event is recorded for each event that occurs in a county with no warning. One unverified warning is counted for each warned county that does not experience an event.

Quality Assurance Rules

In an attempt to reduce the impact of erroneous short-fused warnings on customers and, at the same time, more accurately measure the quality of NWS warnings, the OCWWS Performance Branch has developed a set of rules stating how these short-fused warnings are archived.

• Rule 1: How Warnings are Entered into the Database. All data imported into the warning database are taken directly from the warning. No data are entered into the database from any information other than that represented by the bold-faced parts of the warning sample in Table D.1. Based on this rule, products issued with the improper coding may or may not be imported into the database. Several examples appear on the NWS Verification Web Page.

• Rule 2: Quality Assurance of Overlapping Warnings. When two warnings for a given county overlap in time, the portion of the earlier warning that overlaps the second warning is removed. The expiration time of the first warning is changed to one minute before the issuance time of the second warning. Several examples appear on the NWS Verification Web Page.

Lead Time

For verification purposes, the definition of the term "event" is given above. The lead time for each flash flood event is computed separately for each county by subtracting the time of warning issuance from the time

TABLE D.1 Flash Flood Warning Sample

```
WGUS51 KWSH 010000                    <——— WMO PRODUCT ID
FFWWSH                                <——— WARNING TYPE and WFO
DCC001-003-005-010200-                <——— COUNTY and STATE WARNED

BULLETIN - EAS ACTIVATION REQUESTED
FLASH FLOOD WARNING
NATIONAL WEATHER SERVICE SILVER SPRING MD
700 PM EST WED JAN 1 2001             <——— DATE AND ISSUANCE TIME

THE NATIONAL WEATHER SERVICE IN SILVER SPRING HAS ISSUED A

* FLASH FLOOD WARNING FOR...
  WASHINGTON COUNTY IN THE DISTRICT OF COLUMBIA
  JEFFERSON COUNTY IN THE DISTRICT OF COLUMBIA
  REAGAN COUNTY IN THE DISTRICT OF COLUMBIA

* UNTIL 900 PM EST                    <——— EXPIRATION TIME

* AT 700 PM EST...DOPPLER RADAR INDICATED A THUNDERSTORM
PRODUCING HEAVY RAINS 2 MILES WEST OF ADAMS MORGAN...MOVING
EAST AT 15 MPH.

EXCESSIVE RUNOFF WILL CAUSE FLOODING OF LOW LYING AREAS
ANDINTERSECTIONS. POOR DRAINAGE AREAS ARE MOST AT RISK. DO
NOT DRIVE THROUGH FLOODED ROADS.

LAT...LON 3778 9752 3748 9752 3749 9724 3785 9724

$$
```

when the event first occurred in the county. Negative values are converted to zero. If one or more events occur in a county with no warning in effect, each unwarned event is assigned a lead time of zero. Average lead time is computed from all lead times listed in the event database, including zeroes. The percentage of events with lead time greater than zero is also computed.

Display of Verification Statistics

NWS employees access FFW verification statistics through the Stats on Demand feature of the NWS Verification Web Page. Stats on Demand accesses an interactive database that provides verification statistics customized to the user's request. The user may request data by:

TABLE D.2 Example of Format for Reporting Preliminary Flash Flood Warning Verification Statistics

Preliminary Verification Statistics	Flash floods
Number of warnings issued	
Number of verified warnings	
Number of unverified warnings	
Number of events	
Number of events with warnings	
Number of events without warnings	
Average lead time	
Probability of detection	
False alarm rate	
Critical success index	

- one or more dates (select beginning and ending date);
- one or more counties, WFOs, states, NWS regions, or the contiguous United States.

Backup Mode for Warnings

When a WFO goes into backup mode, FFWs are still sorted by county, so all FFWs issued by the backup office are attributed to the primary WFO.

Preliminary Flash Flood Warning Reports

The regional headquarters will report to the OCWWS Performance Branch no later than the 14th of each month preliminary flash flood warning verification statistics for the previous month. When the 14th of the month falls on a weekend or holiday, the statistics are due the last business day prior to the 14th. Present the statistics in the format given in Table D.2. The OCWWS Performance Branch subsequently collates the regional data into a national summary.

Appendix E

Acronyms

AFB	Air Force Base
AFWA	Air Force Weather Agency
AFWS	Automated Flood Warning Systems
AGL	above ground level
AHPS	Advanced Hydrologic Prediction Services (NWS)
ALERT	Automated Local Evaluation in Real Time
AMBER	Areal Mean Basin Estimated Rainfall
ANC	Auto-Nowcast System
AP	anomalous propagation
ARSR	Air Route Surveillance Radar (FAA)
ASR	Airport Surveillance Radar (FAA)
AWIPS	Advanced Weather Interactive Processing System (NWS)
BASC	Board on Atmospheric Sciences and Climate
BDDS	Base Data Distribution System
CA OES	California Office of Emergency Services
CAPS	Center for Analysis and Prediction of Storms
CIWS	Corridor Integrated Weather System (FAA)
CODE	Common Operations and Development Environment
CONUS	continental United States
CRAFT	Collaborative Radar Acquisition Field Test
CWA	county warning area
CWI	commercial weather industry
DEM	digital elevation model
DOC	Department of Commerce
DoD	Department of Defense
DOT	Department of Transportation

DSD	drop size distribution
EAS	Emergency Activation System
ENSO	El Niño-Southern Oscillation
ETA	NCEP numerical weather prediction model
FAA	Federal Aviation Administration
FAR	false alarm ratio
FFMP	Flash Flood Monitoring and Prediction
FFW	flash flood warning
FSL	Forecast Systems Laboratory
GAO	General Accounting Office
GIS	Geographical Information Systems
GIUH	geomorphological instantaneous unit hydrograph
HPC	Hydrometeorological Prediction Center
IFPS	Interactive Forecast Preparation System
KEYX	Edwards Air Force Base NEXRAD
KNYX	San Diego NEXRAD
KSOX	Santa Ana Mountains NEXRAD
KVBX	Vandenberg Air Force Base NEXRAD
KVTX	Sulphur Mountain NEXRAD
LAX	Los Angeles International Airport
LLJ	low-level jet
LOX	Los Angeles-Oxnard Weather Forecast Office
LST	local standard time
MAR	modernization and associated restructuring
MCS	mesoscale convective systems
MM5	Pennsylvania State University-National Center for Atmospheric Research Mesoscale Model
MSL	above mean sea level
MTR	San Francisco Weather Forecast Office
NAS	National Academy of Sciences
NCAR	National Center for Atmospheric Research

NCDC	National Climatic Data Center
NCEP	National Centers for Environmental Prediction
NCWF	National Convective Weather Forecast
NDFD	National Digital Forecast Database
NEXRAD	Next Generation Weather Radar
NMOC	National Meteorology & Oceanography Command
NOAA	National Oceanic and Atmospheric Administration
NPI	NEXRAD Product Improvement
NRC	National Research Council
NRL	Naval Research Laboratory
NWP	numerical weather prediction
NWS	National Weather Service
OCWWS	Office of Climate, Water, and Weather Services (NOAA)
ORDA	open radar data acquisition
ORPG	open radar product generator
POD	probability of detection
PPS	Precipitation Processing System (NEXRAD)
PRF	pulse-repetition frequency
PUP	principal user processor
PW	precipitable water
QPE	quantitative precipitation estimation
QPF	Quantitative Precipitation Forecasting
RASS	radio acoustic sounding systems
RDA	radar data acquisition
RFC	river forecast center
RPG	radar product generator
SGX	San Diego Weather Forecast Office (FAA)
TDWR	Terminal Doppler Weather Radar
TMU	traffic management units
UTC	coordinated universal time
VCP	volume coverage patterns
VCWPD	Ventura County Watershed Protection District

VDRAS	Variational Doppler Radar Analysis System
VPR	vertical profile of reflectivity
WFO	Weather Forecast Office (NWS)
WRF	Weather Research and Forecasting
WSR-74C	Weather Surveillance Radar-1974 C-band
WSR-88D	Weather Surveillance Radar-1988 Doppler

Appendix F

Community Participation

The committee would like to express its sincere appreciation to the following individuals for providing valuable information and discussion throughout this study:

Steve Bennet, Supervisor of the First District of Ventura County, California
Edward Berkowitz, National Oceanic and Atmospheric Administration
David Curtis, One Rain, Inc.
David Danielson, National Oceanic and Atmospheric Administration
Larry Dunn, National Oceanic and Atmospheric Administration
Charles Fisk, Naval Air Warfare Center
Alan Fox, Fox Weather, LLC
Joe Friday, American Meteorological Society
Konstantine Georgakakos, Hydrologic Research Center
Dale Givner, Resident of Ventura County, California
Thomas Graziano, National Oceanic and Atmospheric Administration
Bill and Tina Kee, Landowners of the Sulphur Mountain Radar Site
Daniel Keeton, National Oceanic and Atmospheric Administration
Frank Kelly, National Oceanic and Atmospheric Administration
Milton Kramer, Resident of Ventura County, California
Jayme Laber, Ventura County Watershed Protection District
Kathy Long, Supervisor of the Third District of Ventura County, California
Jim Manitakos, SRI International
Timothy McClung, National Oceanic and Atmospheric Administration
Daniel Melendez, National Oceanic and Atmospheric Administration
Todd Morris, National Oceanic and Atmospheric Administration
Jack Paris, DigitalGlobe, Inc.
Richard Paulus, Space and Naval Warfare Systems Center
Juris Petriceks, SRI International (retired)
F. Martin Ralph, National Oceanic and Atmospheric Administration

Robert Saffle, Mitretek Systems
John Sokich, National Oceanic and Atmospheric Administration
Gary Strickland, National Oceanic and Atmospheric Administration
Paula Sturdevant-Rees, University of Massachusetts
Dolores Taylor, Automated Local Evaluation in Real Time (ALERT) Users Group and National Hydrologic Warning Council

Appendix G

Committee and Staff Biographies

Paul L. Smith (*chair*) is a past director (1981–1996) of the Institute of Atmospheric Sciences of the South Dakota School of Mines and Technology and is now a professor emeritus. He received his Ph.D. in electrical engineering from Carnegie Institute of Technology (now Carnegie-Mellon University). He served as chief scientist at Air Weather Service (AWS) Headquarters, Scott Air Force Base, during 1974–1975 and received the Award for Meritorious Civilian Service for his contributions to the AWS radar program. He served on the Executive Committee of the International Commission on Clouds and Precipitation from 1988 to 1996, as director of the South Dakota Space Grant Consortium from 1991 to 1996, as a member of the National Research Council's National Weather Service (NWS) Modernization Committee from 1997 to 1999, and as chair of the NRC Committee on Weather Radar Technology beyond NEXRAD from 2001 to 2002. He currently serves on the NEXRAD Technical Advisory Committee. His major research interests are in radar meteorology, cloud physics, and weather modification. He manages the armored T-28 research aircraft facility and has worked on the development of various types of meteorological instrumentation. Dr. Smith is a fellow of the American Meteorological Society (AMS) and chaired the AMS Committee on Radar Meteorology on two separate occasions. He received the 1992 Editor's Award from the AMS *Journal of Applied Meteorology*. He is a member of the Weather Modification Association (receiving its 1995 Thunderbird Award), a Life Senior Member of the Institute of Electrical and Electronics Engineers (IEEE), and a member of Sigma Xi.

Ana P. Barros is a professor at Duke University and a visiting scholar at Harvard University. Her research interest is in investigating the dynamics of water presence and water pathways in the environment. The goal is to improve understanding of the physics of the hydrological cycle at all spatial and temporal scales, and to apply this new knowledge to research and develop technologies for environmental assessment, prediction and control. Barros's work is interdisciplinary and addresses fundamental engineering science questions in the areas of climate, hydrometeorology, geomorphology, ecology, hydraulics and hydrology, and their linkages to the environmental engineering sciences. Her research activities are conducted using computer models, signal processing and exploratory data analysis, and laboratory and field experiments. She previously served on the NRC Committee on U.S. Geological Survey (USGS) Water Resources Research, 1997–2000.

V. Chandrasekar is a professor in the Electrical Engineering Department at Colorado State University (CSU), where he received his Ph.D. Dr. Chandrasekar has been involved with weather radar systems for more than 20 years and has about 25 years of experience in radar systems. Dr. Chandrasekar has played a key role in developing the CSU-CHILL radar as one of the most advanced meteorological radar systems available for research, and he continues to work actively with the CSU-CHILL radar supporting its research and education mission. He specializes in developing new radar technologies and techniques for solving meteorological problems and has actively pursued applications of polarimetry for cloud microphysical applications, as well as neural network-based radar rainfall estimates and fuzzy logic systems for hydrometeor identification. Dr. Chandrasekar is an avid experimentalist conducting experiments to collect in situ observations to verify the new techniques and technologies. He is coauthor of two textbooks, *Polarimetric Doppler Weather Radar: Principles and Applications* (Cambridge University Press, 2001) and *Introduction to Probability and Random Processes* (McGraw Hill, 1997). He is a fellow of the IEEE and recipient of the National Aeronautics and Space Administration (NASA) technical achievement award.

Gregory S. Forbes earned a Ph.D. from the University of Chicago where he studied tornadoes and severe thunderstorms under Dr. Theodore Fujita—world famous for his invention of the F-scale used to rate tornadoes. From 1978 to 1999, Forbes was a member of the meteorology faculty at Pennsylvania State University and taught courses in weather analysis, forecasting, hydrometeorology, and natural disasters. He has been involved in numerous field measurement programs that included Doppler radars and was lead forecaster for some of these projects. Forbes has used NEXRAD, research multiparameter, and phased-array radar data to evaluate various weather situations. In June 1999, he joined the operational forecasting team at the Weather Channel as its severe weather expert. Forbes is a fellow of the AMS and a present or past member of AMS committees on severe local storms and weather analysis and forecasting.

Eve Gruntfest is professor of geography and environmental studies at the University of Colorado, Colorado Springs. She has been working in the field of natural hazard mitigation for more than 25 years, has published widely, and is an internationally recognized expert in the specialty areas of warning system development and flash flooding. Gruntfest is coeditor of *Coping with Flash Floods,* which brings together papers from leading experts who participated at the 1999 NATO Advanced Studies Institute that she organized and held in Ravello, Italy. She has participated in numerous workshops sharing lessons from research on warning systems and flash flooding. Her current research project aims to change how we warn for short fuse events including flash floods and tornadoes to account for new data sources, new technologies, and new urban demographies.

Witold F. Krajewski is the Rose & Joseph Summers Professor of Water Resources Engineering at the University of Iowa. He received his Ph.D. from Warsaw University of Technology, Poland, in environmental engineering and water resources systems. He was a research hydrologist at the Office of Hydrology of the National Weather Service until 1987 when he joined the University of Iowa. His scientific interests concern multiple aspects of rainfall measuring, modeling, forecasting, and estimation using radar and satellite remote sensing. His current research focuses on modeling uncertainty of multisensor rainfall estimation at a range of temporal and spatial scales. His other work includes optimal estimation and control of water resources and environmental systems. He has published nearly 100 papers in refereed journals. He is a Fellow of the American Meteorological Society and the American Geophysical Union. He has served on numerous committees and panels of these and other professional organizations and on the editorial boards of several journals. He is a coeditor of *Advances in Water Resources.*

Thomas D. Potter is professor of meteorology, emeritus at the University of Utah. He earned his Ph.D. in 1962 from Pennsylvania State University. He spent 23 years in the Air Weather Service, retiring in 1974 as the vice commander of a 10,000 person organization. He then joined the faculty at St. Louis University for 2 years, teaching and doing research in meteorology. From there he became director of the NOAA National Climate Center (now the National Climatic Data Center [NCDC]) in Asheville, North Carolina, for 2 years. Potter was asked to transfer to Washington, D.C., as deputy director and later director of the Environmental Data and Information Service. In 1982 he moved to Geneva as the first director of the World Meteorological Organization (WMO) World Climate Program and later became director of the World Weather Watch Department. Returning to the United States in 1989, he became regional director of the NWS Western Region, implementing the NWS modernization, bringing in new technology and training his people to use it effectively. In 1998, he retired from NWS and joined the University of Utah Meteorology Department as a professor, director of the Cooperative Institute for Regional Prediction, and the leader of the Weather Support Unit for the 2002 Olympics held in Salt Lake City, Utah. Potter is a founding member of Atmospheric Science Advisors, a consulting firm. He is also coeditor of the recent Wiley *Handbook of Weather, Climate, and Water*, and fellow of the AMS, American Indian Science and Engineering Society, and Pennsylvania State University and a member of Sigma Xi.

Rita Roberts is a project scientist with the Research Applications Program (RAP) at the National Center for Atmospheric Research. Rita has extensive experience in analyzing Doppler and polarimetric radar data as well experience in radar siting. She also has used NEXRAD data for flash flood forecasting. Rita is the incoming chairperson for the AMS Committee on Radar Meteorology. In addition to radar meteorology, her other interests include tornado research and thunderstorm nowcasting. Before joining RAP, Roberts worked on the Joint Airport Weather Studies Project, which focused on microbursts and wind shear. She received an M.S. in atmospheric science from Colorado State University in 1998.

Matthias Steiner is a senior research scientist affiliated with the Department of Civil and Environmental Engineering at Princeton University. He received his Ph.D. in environmental sciences (with emphasis in atmospheric science) from the Swiss Federal Institute of Technology (ETH) in Zurich. Dr. Steiner's research interests reach across hydrometeorology, cloud and precipitation physics, mountain meteorology, and radar and satellite meteorology. He is intrigued by the variability of precipitation in space and time and how to measure precipitation with in situ as well as remote sensing instruments. His recent work is focused on understanding the effect of atmospheric moisture on the flow of air within and over complex terrain, and the associated cloud and precipitation processes. Dr. Steiner has been contributing to several interdisciplinary, national, and international field experiments and programs, such as the Mesoscale Alpine Program (MAP), the Tropical Rainfall Measuring Mission (TRMM), and the Tropical Ocean Global Atmosphere (TOGA) Coupled Ocean-Atmosphere Response Experiment (COARE). His work has been published in the leading journals of major professional societies on three continents. Dr. Steiner just completed serving two terms on the AMS Committee on Radar Meteorology. At present, he chairs the Technical Committee on Precipitation of the American Geophysical Union (AGU) Hydrology Section, and he is a member of the Precipitation Missions Science Team of NASA. He is a fellow of the Royal Meteorological Society and was the recipient of the 2002 Editor's Award for the AMS *Journal of Hydrometeorology*.

Roger M. Wakimoto is a professor and the former chair of the Department of Atmospheric Sciences at the University of California, Los Angeles. He received his Ph.D. in geophysical science from the University of Chicago. His professional interests include research and field work in mesometeorology with particular emphasis in severe convective phenomena. Dr. Wakimoto has been a part of or co-principal investigator for numerous field experiments, including VORTEX (Verification of the Origins of Rotation in Tornadoes Experiment) and FASTEX (Fronts and Atlantic Storm Track Experiment). Presently he serves on the NOAA National Centers for Environmental Prediction Advisory Panel and the AMS Committee on Radar Meteorology. He is also an associate editor of *Monthly Weather Review*, and a member of the Board on Atmospheric Sciences and Climate.

Staff

Julie Demuth is a program officer for the Board on Atmospheric Sciences and Climate (BASC). She received her B.S. in meteorology from the University of Nebraska-Lincoln and her M.S. in atmospheric science from Colorado State University. Her master's research focused on developing techniques for objectively estimating the intensity and wind structure of tropical cyclones in the Atlantic and East Pacific basins using microwave sounding data. The intensity estimation algorithm is now being run operationally by NOAA's National Hurricane Center during the tropical season. Since joining BASC in March 2003, Julie has worked on studies involving atmospheric dispersion of hazardous materials, weather modification, and road weather research.

Elizabeth A. Galinis is a senior program assistant for the Board of Atmospheric Sciences and Climate. She received her B.S. in marine science from the University of South Carolina in 2001. Since her start at the National

Academies in March 2002, she has worked on studies involving NEXRAD weather radar, weather modification, climate sensitivity, climate change, and radiative forcings. Liz is currently in graduate school at Johns Hopkins University, pursuing her master's in environmental science and policy.